# REVISE EDEXCEL
# A2 Mathematics
## C3 C4

# REVISION GUIDE

Series Consultant: Harry Smith

Author: Harry Smith

## A note from the publisher

In order to ensure that this resource offers high-quality support for the associated Pearson qualification, it has been through a review process by the awarding body. This process confirms that this resource fully covers the teaching and learning content of the specification or part of a specification at which it is aimed. It also confirms that it demonstrates an appropriate balance between the development of subject skills, knowledge and understanding, in addition to preparation for assessment.

Endorsement does not cover any guidance on assessment activities or processes (e.g. practice questions or advice on how to answer assessment questions), included in the resource nor does it prescribe any particular approach to the teaching or delivery of a related course.

While the publishers have made every attempt to ensure that advice on the qualification and its assessment is accurate, the official specification and associated assessment guidance materials are the only authoritative source of information and should always be referred to for definitive guidance.

Pearson examiners have not contributed to any sections in this resource relevant to examination papers for which they have responsibility.

Examiners will not use endorsed resources as a source of material for any assessment set by Pearson.

Endorsement of a resource does not mean that the resource is required to achieve this Pearson qualification, nor does it mean that it is the only suitable material available to support the qualification, and any resource lists produced by the awarding body shall include this and other appropriate resources.

ALWAYS LEARNING

**PEARSON**

# Contents

. . . . . . . . . . . . . . . . . . . . . . . . . . . . . . . . . . . . .

**A small bit of small print**

Edexcel publishes Sample Assessment Material and the Specification on its website. This is the official content and this book should be used in conjunction with it. The questions in *Now try this* have been written to help you practise every topic in the book. Remember: the real exam questions may not look like this.

# Algebraic fractions

You will usually be able to simplify algebraic fractions in your C3 exam using these steps:

**1 Factorise...**
...expressions in the numerators or denominators into linear factors.

**2 Cancel...**
...common factors in the numerator and denominator of each fraction.

**3 Add or subtract...**
...any fractions by finding a common denominator, to make a SINGLE FRACTION.

If you FACTORISE and CANCEL first then your fractions will be easier to ADD OR SUBTRACT.

You might need to do steps 1 and 2 again once you have added or subtracted your fractions.

## Common denominators

To add or subtract algebraic fractions with different denominators you need to find a common denominator. Here are two examples:

**1** $\dfrac{3}{x} + \dfrac{5}{2x + 1} = \dfrac{3(2x + 1)}{x(2x + 1)} + \dfrac{5x}{x(2x + 1)} = \dfrac{3(2x + 1) + 5x}{x(2x + 1)}$

The common denominator is the PRODUCT of the two denominators.

**2** $\dfrac{2}{x + 1} - \dfrac{2x}{(x - 2)(x + 1)} = \dfrac{2(x - 2)}{(x + 1)(x - 2)} - \dfrac{2x}{(x + 1)(x - 2)} = \dfrac{2(x - 2) - 2x}{(x + 1)(x - 2)}$

Once you are confident you might be able to skip the step shown in green above.

The denominators already SHARE a factor of $(x + 1)$, so you only have to change the first fraction.

After this step, you can simplify the fractions further by expanding the brackets in the numerator and collecting like terms.

## Worked example

Express $\dfrac{2(3x + 2)}{9x^2 - 4} - \dfrac{2}{3x + 1}$

as a single fraction in its simplest form.   **(4 marks)**

$\dfrac{2(3x + 2)}{9x^2 - 4} - \dfrac{2}{3x + 1} = \dfrac{2\cancel{(3x + 2)}}{\cancel{(3x + 2)}(3x - 2)} - \dfrac{2}{3x + 1}$

$= \dfrac{2}{3x - 2} - \dfrac{2}{3x + 1}$

$= \dfrac{2(3x + 1) - 2(3x - 2)}{(3x - 2)(3x + 1)}$

$= \dfrac{6}{(3x - 2)(3x + 1)}$

You will probably have to do **multiple steps** so get used to writing out your working neatly. You can show which factors you are **cancelling** by drawing a neat line through them.

After you have added or subtracted your fractions you should expand the brackets in the numerator and simplify again:

$2(3x + 1) - 2(3x - 2) = \cancel{6x} + 2 - \cancel{6x} + 4$
$= 6$

You can leave the final denominator factorised like this, or multiply it out.

## Now try this

1. Simplify fully $\dfrac{3x^2 - 8x - 3}{x^2 - 9}$   **(3 marks)**

2. Express $\dfrac{x + 5}{2x^2 + 7x - 4} - \dfrac{1}{2x - 1}$

   as a single fraction in its simplest form.   **(4 marks)**

Watch out for the difference of two squares.
Use $a^2 - b^2 = (a + b)(a - b)$

# Algebraic division

You might need to find missing coefficients when a CUBIC or QUARTIC expression is divided by a QUADRATIC expression. You can use LONG DIVISION, but make sure you set your work out neatly.

Here is the working for $\dfrac{3x^4 - 6x^3 + x - 2}{x^2 - 1}$:

You need to multiply $(x^2 - 1)$ by $3x^2$ to get the term $3x^4$, so the first term in your answer is $3x^2$

The $x$ coefficient is 0, so write $+ 0x$

The $x^2$ coefficient is 0, so write $+ 0x^2$

$$
\begin{array}{r}
3x^2 - 6x + 3 \phantom{00000} \\
x^2 + 0x - 1 \,\overline{\big)\,3x^4 - 6x^3 + 0x^2 + \phantom{0}x - 2} \\
3x^4 + 0x^3 - 3x^2 \phantom{000000000} \\
\hline
-6x^3 + 3x^2 + \phantom{0}x - 2 \\
-6x^3 + 0x^2 + 6x \phantom{0000} \\
\hline
3x^2 - 5x - 2 \\
3x^2 + 0x - 3 \\
\hline
-5x + 1
\end{array}
$$

$3x^2 \times (x^2 + 0x - 1) = 3x^4 + 0x^3 - 3x^2$

Always line up terms with the same power of $x$.

Be careful with negative terms when subtracting: $-2 - (-3) = 1$

If you are dividing by a quadratic, the remainder will be a LINEAR term.

So $\dfrac{3x^4 - 6x^3 + x - 2}{x^2 - 1} = (3x^2 - 6x + 3) + \dfrac{(-5x + 1)}{x^2 - 1}$

Quotient — $(3x^2 - 6x + 3)$

Divisor — $x^2 - 1$

## Worked example

Given that
$$\frac{3x^4 - 2x^3 - 5x^2 - 4}{x^2 - 4} \equiv ax^2 + bx + c + \frac{dx + e}{x^2 - 4}, \quad x \neq \pm 2$$
find the values of the constants $a$, $b$, $c$, $d$ and $e$. **(4 marks)**

$3x^4 - 2x^3 - 5x^2 - 4 \equiv (ax^2 + bx + c)(x^2 - 4) + dx + e$
$\equiv ax^4 + bx^3 + cx^2 - 4ax^2 - 4bx - 4c + dx + e$
$\equiv ax^4 + bx^3 + (c - 4a)x^2 + (d - 4b)x + (e - 4c)$

$x^4$ terms $\rightarrow \underline{a = 3}$      $x$ terms $\rightarrow d - 4b = 0$

$x^3$ terms $\rightarrow \underline{b = -2}$         $d + 8 = 0$

$x^2$ terms $\rightarrow c - 4a = -5$         $\underline{d = -8}$

$\qquad c - 12 = -5$      Constant terms $\rightarrow e - 4c = -4$

$\qquad\quad \underline{c = 7}$         $e - 28 = -4$

$\qquad\qquad\qquad\qquad\qquad \underline{e = 24}$

You can also **compare coefficients** to find the missing coefficients. Follow these steps:

1. Multiply both sides by the divisor.

2. Expand the brackets **carefully** then collect like terms.

3. Compare coefficients on each side, starting with the highest power of $x$.

As long as you write down what each constant is equal to, you don't need to write out the whole expression at the end.

## Now try this

Given that
$$\frac{2x^4 + 4x^2 - x + 2}{x^2 - 1} \equiv ax^2 + bx + c + \frac{dx + e}{x^2 - 1}, \quad x \neq \pm 1$$
find the values of the constants $a$, $b$, $c$, $d$ and $e$.
**(4 marks)**

Whichever method you use, make sure you either:
• write out the expression in full with the constants in place, or
• write $a = ...$, $b = ...$, etc.

# Functions

A function maps numbers in its DOMAIN onto numbers in its RANGE. Here is an example:

f is the NAME of the function. You can use any letter, but f and g are the most common.

$$f(x) = \sqrt{x - 2}, \qquad x \geqslant 2$$

$x$ is the INPUT. You say 'f of $x$'. You can also write $f : x \to \sqrt{x - 2}$ and say 'f maps $x$ onto $\sqrt{x - 2}$'.

This tells you what the function does to $x$.
$$f(18) = \sqrt{18 - 2}$$
$$= \sqrt{16} = 4$$

This is the DOMAIN of the function. The function is only defined for these INPUT values. The RANGE of this function is $f(x) \geqslant 0$. This tells you all the possible OUTPUT values for the function.

## Composite functions

If you apply two functions one after the other, you can write a SINGLE FUNCTION which has the same effect as the two combined functions. This is called a COMPOSITE FUNCTION.

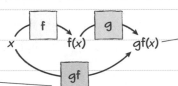

The function gf has the same effect as applying function f THEN applying function g.

The ORDER is important. The function being applied FIRST goes CLOSEST to the $x$.

The functions f and g are defined by
$$f : x \mapsto 1 + 4x^3, \quad x \in \mathbb{R}$$
$$g : x \mapsto 2 - \frac{1}{x}, \quad x \in \mathbb{R}, \quad 0 < x \leqslant 4$$

(a) Show that the composite function gf is
$$gf : x \mapsto \frac{1 + 8x^3}{1 + 4x^3} \qquad \textbf{(4 marks)}$$

$$gf(x) = 2 - \frac{1}{1 + 4x^3}$$
$$= \frac{2(1 + 4x^3) - 1}{1 + 4x^3}$$
$$= \frac{1 + 8x^3}{1 + 4x^3}$$

(b) Solve $gf(x) = 0$ $\qquad$ **(2 marks)**

$$\frac{1 + 8x^3}{1 + 4x^3} = 0 \quad \text{so} \quad 1 + 8x^3 = 0$$
$$8x^3 = -1$$
$$x^3 = -\frac{1}{8}$$
$$x = -\frac{1}{2}$$

gf($x$) means you do f first, then g. To find an expression for gf($x$) you need to substitute the **whole expression** for f($x$) into the expression for g($x$):
$$g : x \mapsto 2 - \frac{1}{x}$$

Substitute the whole expression for f($x$) here.

## Real numbers

The symbol $\mathbb{R}$ means all the REAL NUMBERS. These are all the positive and negative numbers, including 0. Here are some domains of functions:

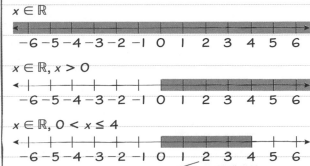

$x \in \mathbb{R}$

$x \in \mathbb{R}, x > 0$

$x \in \mathbb{R}, 0 < x \leqslant 4$

You can write this as (0, 4].
The square bracket shows that 4 is included.

The functions f and g are defined by
$$f : x \mapsto 3x + 2, \quad x \in \mathbb{R}$$
$$g : x \mapsto \frac{x}{x + 2}, \quad x \in \mathbb{R}, \quad x \neq -2$$

Don't worry about the domain. If you need to state a domain you will be told to do so in the question.

(a) Find gf(−2). $\qquad$ **(2 marks)**

(b) Show that the composite function fg is
$$fg : x \mapsto \frac{5x + 4}{x + 2} \qquad \textbf{(4 marks)}$$

(c) Solve the equation $f(x) = [f(x)]^2$ $\qquad$ **(4 marks)**

# Graphs and range

You can use graphs of the form $y = f(x)$ or $y = g(x)$ to represent functions. The $y$-coordinates tell you the output values of the function (the RANGE).

## Worked example

The function f has domain $-5 \leqslant x \leqslant 7$. A sketch of the graph of $y = f(x)$ is shown below.

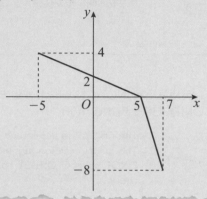

The range is the possible output values of the function. This is all the $y$-coordinates on the graph of the function:

Use the same type of inequality as the domain.

$$-8 \leqslant f(x) \leqslant 4$$

Use f(x) or $y$ as the variable.
Do not use $x$ when writing the range.

(a) Write down the range of f. **(1 mark)**

$-8 \leqslant f(x) \leqslant 4$

(b) Find ff(5). **(2 marks)**

ff(5) = f(O) = 2

ff(5) means f[f(5)]
When $x = 5$, $y = 0$, so f(5) = 0
When $x = 0$, $y = 2$, so f(O) = 2

Work out the value of the function near the edges of the domain. If you have a **graphing calculator** you could plot the graph to get an idea of its shape.

This is a quadratic function. You can find the minimum value by completing the square. Look at the edges of the domain to find the maximum.

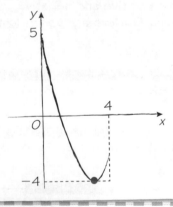

## Worked example

Find the range of the functions defined by:

(a) $f(x) = \dfrac{x + 1}{x - 3}, \quad x > 5$ **(2 marks)**

When $x = 5$, $\dfrac{x + 1}{x - 3} = \dfrac{6}{2} = 3$

As $x \to \infty$, $\dfrac{x + 1}{x - 3} \to 1$ but is always > 1.

Range is $1 < f(x) < 3$

(b) $g(x) = x^2 - 6x + 5, \quad x \in \mathbb{R}, \quad 0 \leqslant x \leqslant 4$

$g(x) = (x - 3)^2 - 4$ **(2 marks)**

So $g(x)$ has a minimum at $(3, -4)$.

Maximum value of $g(x)$ occurs at $x = 0$.

$g(O) = 0^2 - 6 \times 0 + 5 = 5$

Range is $-4 \leqslant g(x) \leqslant 5$

## Now try this

1. The function f has domain $-5 \leqslant x \leqslant 8$ and is linear from $(-5, -6)$ to $(0, 4)$ and from $(0, 4)$ to $(8, 0)$.

(a) Write down the range of f. **(1 mark)**

(b) Find ff(−1). **(2 marks)**

2. The function g is defined by
$$g : x \mapsto \dfrac{3x - 7}{x^2 - 5x + 6} - \dfrac{2}{x - 3}, \quad x \in \mathbb{R}, \quad x > 3$$

(a) Show that $g(x) = \dfrac{1}{x - 2}, \quad x > 3$ **(3 marks)**

(b) Find the range of g. **(2 marks)**

You need to use the fact that the two sections are linear to work out the values of the function between the given points.

# Inverse functions

For a function f, the INVERSE of f is the function that UNDOES f. You write the inverse as $f^{-1}$. If you apply f then $f^{-1}$, you will end up back where you started.

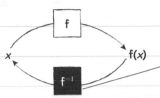

If you apply f, then $f^{-1}$ you have applied the COMPOSITE FUNCTION $f^{-1}f$. The output of $f^{-1}f$ is the SAME as the input. You can write:
$$f^{-1}f(x) = ff^{-1}(x) = x$$

## Finding the inverse

To find the inverse of a function given in the form $f(x) = \ldots$ you need to:

**1** Write the function in the form $y = \ldots$

**2** Rearrange to make $x$ the subject

**3** Swap any $y$'s for $x$'s and rewrite as $f^{-1}(x) = \ldots$

You aren't asked to state the domain here, so you don't need to include it in your answer.

## Worked example

The function f is defined by:
$$f : x \mapsto \frac{2x + 1}{4 - x}, \quad x \in \mathbb{R}, \quad x \neq 4$$

Find $f^{-1}(x)$.      **(3 marks)**

$$y = \frac{2x + 1}{4 - x}$$
$$y(4 - x) = 2x + 1$$
$$4y - xy = 2x + 1$$
$$4y - 1 = 2x + xy$$
$$4y - 1 = x(2 + y)$$
$$x = \frac{4y - 1}{2 + y}$$
$$f^{-1}(x) = \frac{4x - 1}{2 + x}$$

## Worked example

The function g is defined by:
$$g : x \mapsto \frac{4}{x + 1}, \quad x \in \mathbb{R}, \quad x > 3$$

(a) Find $g^{-1}(x)$.      **(3 marks)**

$$y = \frac{4}{x + 1}$$
$$y(x + 1) = 4$$
$$xy + y = 4$$
$$xy = 4 - y$$
$$x = \frac{4 - y}{y} \quad \text{so} \quad g^{-1}(x) = \frac{4 - x}{x}$$

(b) Find the domain of $g^{-1}$.      **(2 marks)**

When $x = 3$, $\frac{4}{x + 1} = 1$

As $x \to \infty$, $\frac{4}{x + 1} \to 0$ but is always $> 0$.

So range of g is $0 < g(x) < 1$

So domain of $g^{-1}$ is $0 < x < 1$

## Golden rule

The RANGE of a function is the DOMAIN of its INVERSE, and vice versa.

You can often use this rule to find the domain of an inverse function.

## EXAM ALERT!

Find the range of g to work out the domain of $g^{-1}$. If you have a graphing calculator you can sketch the graph of $y = \frac{4}{x + 1}$ to see what it looks like.

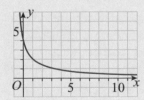

You are interested in the $y$-values for $x > 3$.

Students have struggled with this topic in recent exams – be prepared!

## Now try this

The function h is defined by:
$$h : x \mapsto \frac{x + 5}{x}, \quad x \in \mathbb{R}, \quad x > 1$$

(a) Find $h^{-1}(x)$.      **(3 marks)**

(b) Write down the range of $h^{-1}$.      **(1 mark)**

(c) Find the domain of $h^{-1}$.      **(2 marks)**

Write $y = \frac{x + 5}{x}$ then rearrange to make $x$ the subject. $x$ appears twice on the right-hand side so you will need to factorise to get $x$ on its own.

# Inverse graphs

The graph of $y = f^{-1}(x)$ is the graph of $y = f(x)$ reflected in the line $y = x$.

## Worked example

The diagram shows part of the curve with equation $y = f(x)$. The curve intersects the coordinate axes at $(-5, 0)$ and $(0, 2)$.

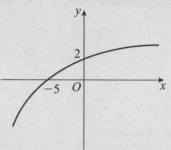

Sketch the curve with equation $y = f^{-1}(x)$.  **(2 marks)**

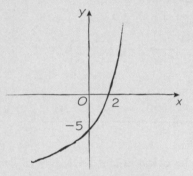

You need to reflect the curve in the line $y = x$.

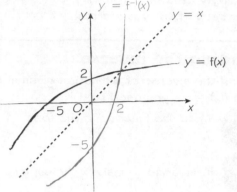

You can reflect a point in the line $y = x$ by swapping the x- and y-coordinates:
$(-5, 0) \rightarrow (0, -5)$ and $(0, 2) \rightarrow (2, 0)$

## Existence of the inverse

The inverse of a function ONLY EXISTS if the function maps each point in its domain to a UNIQUE point in its range. This is sometimes called a ONE-TO-ONE function. If f is a one-to-one function then a horizontal line drawn anywhere on the graph of $y = f(x)$ will never cross the graph more than ONCE.

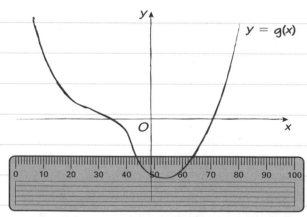

g is NOT a one-to-one function because a horizontal line crosses the graph of $y = g(x)$ more than once.

## Worked example

The function h is defined by
$$h: x \mapsto x^2 + 1, \quad x \in \mathbb{R}$$

Explain why the function h does not have an inverse.  **(1 mark)**

h is not a one-to-one function.

Both of these answers are correct:
• h is not a one-to-one function
• h is a many-to-one function.
You can also use the word 'mapping' instead of 'function' in your answer.

## Now try this

The function f has domain $-8 \leqslant x \leqslant 4$ and is linear from $(-8, -5)$ to $(5, 0)$ and from $(5, 0)$ to $(6, 4)$. A sketch of the graph of $y = f(x)$ is shown.

Sketch the graph of $y = f^{-1}(x)$. Show the coordinates of the points corresponding to $A$ and $B$.  **(3 marks)**

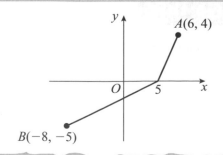

# Modulus

The MODULUS of a number is its positive numerical value. You write the modulus of $x$ as $|x|$.
You can use the graph of $y = f(x)$ to sketch the graphs of $y = |f(x)|$ and $y = f(|x|)$.

$y = f(x)$     **1** $y = |f(x)|$     **2** $y = f(|x|)$

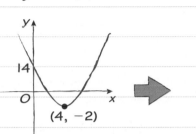

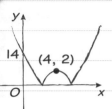

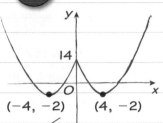

Any points BELOW the $x$-axis are reflected in the $x$-axis. Every point on the curve must have a POSITIVE $y$-coordinate.

Replace the curve to the LEFT of the $y$-axis with a reflection of the curve to the RIGHT of the $y$-axis.

## Worked example

The diagram shows a sketch of the curve with equation $y = f(x)$.

$B(5, 5)$

$A(-3, -2)$

On separate diagrams sketch the following graphs, showing the coordinates of the points corresponding to $A$ and $B$.

(a) $y = |f(x)|$     **(3 marks)**     (b) $y = f(|x|)$     **(3 marks)**

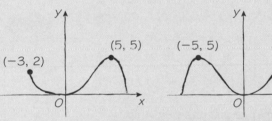

$(-3, 2)$    $(5, 5)$       $(-5, 5)$    $(5, 5)$

## Sketching $y = |ax + b|$

You can sketch the modulus of a linear function by sketching the graph, then reflecting any points that are below the $x$-axis.

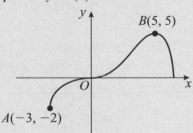

$y = \tfrac{1}{2}x + 1$        $y = |\tfrac{1}{2}x + 1|$

The graph of $y = |ax + b|$ is always a V-shape.

## Worked example

The function f is defined by
$$f : x \to 3|x| - 5, \quad x \in \mathbb{R}$$
State the range of f.     **(2 marks)**

$f(x) \geqslant -5$

$|x|$ is always greater than or equal to 0.

## Now try this

The diagram shows a sketch of the curve with equation $y = f(x)$.

On separate diagrams sketch the following graphs.

(a) $y = |f(x)|$                          **(3 marks)**

(b) $y = f(|x|)$                          **(3 marks)**

In each case, show the coordinates corresponding to the turning points $A$ and $B$.

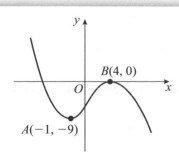

$B(4, 0)$

$A(-1, -9)$

# Transformations of graphs

In your C1 exam you used these transformations of the graph of $y = f(x)$:

- $y = f(x) + a$    Translation $\begin{pmatrix} 0 \\ a \end{pmatrix}$

- $y = f(x + a)$    Translation $\begin{pmatrix} -a \\ 0 \end{pmatrix}$

- $y = af(x)$    Vertical stretch, scale factor $a$

- $y = f(ax)$    Horizontal stretch, scale factor $\frac{1}{a}$

- $y = -f(x)$    Reflection in the $x$-axis

- $y = f(-x)$    Reflection in the $y$-axis.

You need to be able to combine these transformations to sketch more complicated function graphs.

## Golden rule

Carry out transformations in this order:

**1** Anything 'inside' the function brackets

**2** Multiples or modulus of the whole function

**3** Addition or subtraction outside the function brackets.

$$y = \frac{1}{2}f(|x|) + 4$$

       **2**         **1**         **3**

## Worked example

The diagram shows part of the graph of $y = f(x)$, $x \in \mathbb{R}$. The graph consists of two line segments that meet at the point $P(4, -3)$.

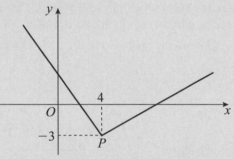

Sketch, on separate diagrams, the graphs of

(a) $y = 2f(x + 4)$ **(3 marks)**    (b) $y = |f(-x)|$ **(3 marks)**

On each diagram, show the coordinates of the point corresponding to $P$.

(a)
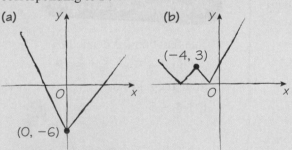

For part (a) you need to carry out a translation $\begin{pmatrix} -4 \\ 0 \end{pmatrix}$ followed by a vertical stretch with scale factor 2.

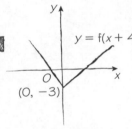 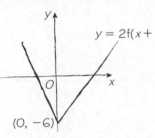

For part (b) you need to carry out a reflection in the $y$-axis followed by a modulus.

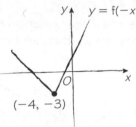 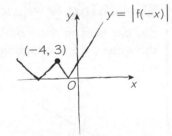

Have a look at page 7 for a reminder about sketching the modulus of a function.

## Now try this

The diagram shows a sketch of $y = f(x)$. The graph has turning points at $P$ and $Q$.

(a) Write down the coordinates of the point to which $Q$ is transformed on the curve with equation
    (i)   $y = 2f(2x)$      (ii)   $y = |f(x + 4)|$        **(4 marks)**

(b) Sketch, on separate diagrams, the graphs of
    (i) $y = f(-x) + 3$      (ii)   $y = -|f(x)|$        **(6 marks)**

Indicate on each diagram the coordinates of any turning points.

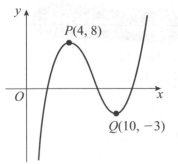

# Modulus equations

Solving an equation involving a modulus function is a bit like solving two equations. You need to consider the situations when the argument (the part inside the modulus) is POSITIVE and NEGATIVE separately. You can use a graph to check that your answers make sense.

## Worked example

The function f is defined by
$$f : x \mapsto |3x - 6|, \quad x \in \mathbb{R}$$

(a) Sketch the graph with equation $y = f(x)$, showing the coordinates of the points where the graph cuts or meets the axes. **(2 marks)**

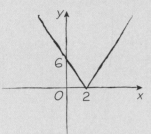

(b) Solve $f(x) = x$ **(3 marks)**

$$3x - 6 = x \qquad\qquad -(3x - 6) = x$$
$$3x = x + 6 \qquad\qquad -3x + 6 = x$$
$$2x = 6 \qquad\qquad 6 = 4x$$
$$x = 3 \qquad\qquad x = \frac{3}{2}$$

## EXAM ALERT!

To solve $|3x - 6| = x$ you need to solve **two** equations:

- Positive argument: $3x - 6 = x$
- Negative argument: $-(3x - 6) = x$

Use your graph to check that the answers definitely exist:

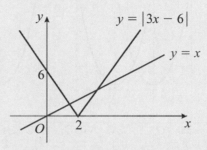

If $y = x$ crosses $y = |3x - 6|$ twice then there are two solutions.

Students have struggled with this topic in recent exams – be prepared!

---

Here is a foolproof way of solving equations involving a modulus:

1. Rearrange the equation so the modulus is on one side.
2. Solve the equation with a positive argument.
3. Solve the equation with a negative argument.
4. Use a graph or plug the answers back into the original equation to check that they exist.

## Worked example

Solve $4 - |x + 2| = \frac{1}{2}x$ **(5 marks)**

$$|x + 2| = 4 - \frac{1}{2}x$$
$$x + 2 = 4 - \frac{1}{2}x \qquad -(x + 2) = 4 - \frac{1}{2}x$$
$$\frac{3}{2}x = 2 \qquad\qquad -x - 2 = 4 - \frac{1}{2}x$$
$$x = \frac{4}{3} \qquad\qquad -6 = \frac{1}{2}x$$
$$x = -12$$

Check: $4 - \left|\frac{4}{3} + 2\right| = 4 - \frac{10}{3} = \frac{2}{3} = \frac{1}{2}\left(\frac{4}{3}\right)$ ✓

$4 - |-12 + 2| = 4 - |-10|$
$$= 4 - 10 = -6 = \frac{1}{2}(-12) ✓$$

## Now try this

1. The function f is defined by
$$f : x \mapsto |2x + 4|, \quad x \in \mathbb{R}$$

(a) Sketch the graph with equation $y = f(x)$, showing the coordinates of the points where the graph cuts or meets the axes. **(2 marks)**

(b) Explain why the equation $f(x) = x$ has no solutions. **(1 mark)**

(c) Solve $f(x) = -x$ **(3 marks)**

2. Solve $2x + 1 = 5 - |x - 1|$ **(5 marks)**

Be careful. This equation has only **one** solution. Find separate solutions for the positive and negative arguments, then plug them both back into the equation to check which one is valid.

# Sec, cosec and cot

Have a quick look at your C2 trigonometry before tackling C3 trigonometry. You should be able to work with radians and exact values of trig functions confidently. You need to learn the names of the RECIPROCALS of sin, cos and tan:

$$\operatorname{cosec}\theta = \frac{1}{\sin\theta} \quad \sec\theta = \frac{1}{\cos\theta} \quad \cot\theta = \frac{1}{\tan\theta}$$

**Golden rule**

Use the THIRD LETTER to remember the reciprocal trig functions:

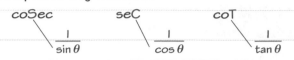

coSec $\quad$ seC $\quad$ coT

$\frac{1}{\sin\theta}$ $\quad$ $\frac{1}{\cos\theta}$ $\quad$ $\frac{1}{\tan\theta}$

## Graphs of sec, cosec and cot

You need to be able to sketch these graphs in your exam. Don't rely on a graphing calculator – learn the SHAPE, ASYMPTOTES and INTERCEPTS for each graph.

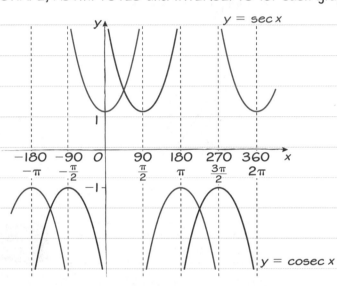

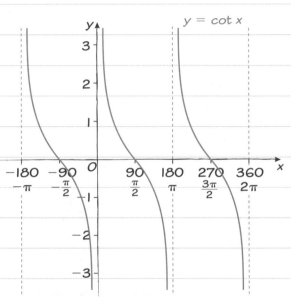

**Worked example**

Sketch the graph of $y = 3\operatorname{cosec}2\theta$ for $0° < \theta < 360°$.

**(2 marks)**

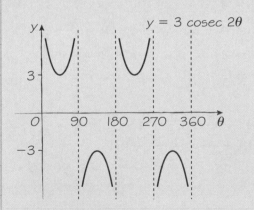

**EXAM ALERT!**

Start with the graph of $y = \operatorname{cosec}\theta$.

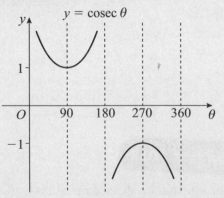

You need to apply two transformations:

- horizontal stretch, scale factor $\frac{1}{2}$
- vertical stretch, scale factor 3.

Students have struggled with this topic in recent exams – be prepared!

**Now try this**

1. Sketch the graph of $y = \cot\left(x - \frac{\pi}{2}\right)$ for $0 \leqslant x \leqslant 2\pi$ **(2 marks)**

2. Sketch the graph of $y = \sec\frac{1}{2}\theta + 1$ for $-360° < \theta < 360°$. **(2 marks)**

Look at page 8 for a reminder about transforming functions.

# Trig equations 1

You need to be able to solve trig equations involving sec, cosec and cot. Make sure you are confident solving C2 trig equations involving sin, cos and tan before revising this topic.

## Multiple solutions

You can use this table to find TWO SOLUTIONS to a trig equation without sketching a graph.

$\alpha$ is the PRINCIPAL VALUE you get from your calculator.

| Function | Radians | Degrees |
|---|---|---|
| $\sin x = k$ | $x = \alpha, \pi - \alpha$ | $x = \alpha, 180° - \alpha$ |
| $\cos x = k$ | $x = \alpha, -\alpha$ | $x = \alpha, -\alpha$ |
| $\tan x = k$ | $x = \alpha, \pi + \alpha$ | $x = \alpha, 180° + \alpha$ |

You can find ALL OTHER SOLUTIONS by adding multiples of $2\pi$ radians, or 360° to these values.

## Worked example

Solve, for $0 \leqslant x \leqslant \pi$,
$$\operatorname{cosec} 3x + 2 = 0$$
Give your answers in terms of $\pi$.          **(5 marks)**

$\operatorname{cosec} 3x = -2$

$\dfrac{1}{\sin 3x} = -2$   so   $\sin 3x = -\dfrac{1}{2}$

$0 \leqslant x \leqslant \pi$,   so   $0 \leqslant 3x \leqslant 3\pi$

$3x = -\dfrac{\pi}{6}, \dfrac{7\pi}{6}, \dfrac{11\pi}{6}, \dfrac{19\pi}{6}, \ldots$

$x = \dfrac{7\pi}{18}, \dfrac{11\pi}{18}$

Rearrange the equation so it is in the form sin ... = ... then solve it in the normal way. Remember to transform the range so you find all the possible solutions.

**Check it!**
You can use your calculator to check your answers.

$$\dfrac{1}{\sin\left(3 \times \dfrac{11\pi}{18}\right)} + 2$$

$$0$$

## EXAM ALERT!

Look out for equations that need to be **factorised**. This equation is the same as

$2A^2 + 7A - 4 = 0$, with $A = \sec x$. When solving quadratic equations involving $\sec x$ or $\operatorname{cosec} x$, both factors won't necessarily give you valid answers. The equations $\sec x = k$ and $\operatorname{cosec} x = k$ have **no solutions** for $-1 < k < 1$, so $\sec x = \frac{1}{2}$ doesn't produce any valid solutions.

The equation $\cot x = k$ does have solutions for any value of $k$, though. Have a look at the graphs on page 10 to see why this is true.

> Students have struggled with this topic in recent exams – be prepared!

## Worked example

Solve, for $-180° \leqslant x \leqslant 180°$,
$$2\sec^2 x + 7\sec x = 4$$
Give your answer in degrees to 1 decimal place.          **(6 marks)**

$2\sec^2 x + 7\sec x - 4 = 0$

$(2\sec x - 1)(\sec x + 4) = 0$

$\sec x = \dfrac{1}{2}$ ✗          $\sec x = -4$ ✓

$\qquad\qquad\qquad\qquad\qquad \dfrac{1}{\cos x} = -4$

$\qquad\qquad\qquad\qquad\qquad \cos x = -\dfrac{1}{4}$

$x = 104.47\ldots°, -104.47\ldots°, 255.52\ldots°$

$x = 104.5°, -104.5°$ (1 d.p.)

## Now try this

1. Solve, for $0 \leqslant x \leqslant 360°$,
$$\cot^2 x + 2 = 3\cot x$$
Give your answers in degrees to 1 decimal place.          **(6 marks)**

2. Solve, for $0 \leqslant x \leqslant \pi$,
$$\sqrt{3}\sec 2x = 2$$
Give your answers in terms of $\pi$.          **(5 marks)**

# Using trig identities

You will need to learn these trigonometric identities for your C3 exam. Make sure you can prove them both, using the C2 identities $\sin^2\theta + \cos^2\theta \equiv 1$ and $\tan\theta \equiv \dfrac{\sin\theta}{\cos\theta}$

**① $\sec^2\theta \equiv 1 + \tan^2\theta$**

$$1 \equiv \sin^2\theta + \cos^2\theta$$
$$\frac{1}{\cos^2\theta} \equiv \frac{\sin^2\theta}{\cos^2\theta} + \frac{\cos^2\theta}{\cos^2\theta}$$
$$\sec^2\theta \equiv \tan^2\theta + 1$$
$$\equiv 1 + \tan^2\theta$$

**② $\operatorname{cosec}^2\theta \equiv 1 + \cot^2\theta$**

$$1 \equiv \sin^2\theta + \cos^2\theta$$
$$\frac{1}{\sin^2\theta} \equiv \frac{\sin^2\theta}{\sin^2\theta} + \frac{\cos^2\theta}{\sin^2\theta}$$
$$\operatorname{cosec}^2\theta \equiv 1 + \frac{1}{\tan^2\theta}$$
$$\equiv 1 + \cot^2\theta$$

If you have to prove an identity you should start with one side, then rearrange it using identities you know until it looks like the other side. It's usually safest to start with the **left-hand side**. Use the fact that $\sec^4 x = (\sec^2 x)^2$, then multiply out the brackets and rearrange. You can use identity 1 above a second time to complete the proof.

## Worked example

Prove that $\sec^4 x - \tan^4 x \equiv \sec^2 x + \tan^2 x$   **(3 marks)**

$$\sec^4 x - \tan^4 x \equiv (1 + \tan^2 x)^2 - \tan^4 x$$
$$\equiv 1 + 2\tan^2 x + \tan^4 x - \tan^4 x$$
$$\equiv 1 + \tan^2 x + \tan^2 x$$
$$\equiv \sec^2 x + \tan^2 x$$

You could also start by writing $(\sec^4 x - \tan^4 x) \equiv (\sec^2 x - \tan^2 x)(\sec^2 x + \tan^2 x)$, then use identity 1 above to show that $\sec^2 x - \tan^2 x \equiv 1$.

## Worked example

Solve, for $0 \leqslant x \leqslant 180°$, the equation
$$\operatorname{cosec}^2 2x - 3\cot 2x = 1$$
Give your answers in degrees to 1 decimal place.

                                         **(7 marks)**

$$1 + \cot^2 2x - 3\cot 2x = 1$$
$$\cot^2 2x - 3\cot 2x = 0$$
$$\cot 2x(\cot 2x - 3) = 0$$

$$\cot 2x = 0 \checkmark \quad \cot 2x = 3 \checkmark$$
$$\tan 2x \to \infty \quad \tan 2x = \frac{1}{3}$$
$$2x = 90°, 270° \quad 2x = 18.43...°, 198.43...°$$
$$x = 45°, 135°, 9.2°, 99.2° \ (1 \text{ d.p.})$$

1. Use $\operatorname{cosec}^2\theta \equiv 1 + \cot^2\theta$ with $\theta = 2x$ to get a **quadratic** equation in $\cot 2x$.
2. **Factorise** the left-hand side to find two values of $\cot 2x$.
3. Use $\cot 2x = \dfrac{1}{\tan 2x}$ to find all the possible values of $2x$.
4. Divide by 2 to find the solutions.

Remember to transform the range:
$$0 \leqslant x \leqslant 180° \text{ so } 0 \leqslant 2x \leqslant 360°$$
You are interested in all the possible values of $2x$ between 0° and 360°.

## Now try this

1. Prove that $\operatorname{cosec}^2 x - \sec^2 x \equiv \cot^2 x - \tan^2 x$
                                      **(3 marks)**

Rearrange identity 1 at the top of the page:
$$\tan^2\theta = \sec^2\theta - 1$$

2. Solve, for $0 \leqslant \theta \leqslant 360°$, the equation
$$3\tan^2\theta + 7\sec\theta = 3$$
Give your answers in degrees to 1 decimal place.
                                      **(6 marks)**

# Arcsin, arccos and arctan

Arcsin, arccos and arctan are the mathematical names of the $\boxed{\sin^{-1}}$, $\boxed{\cos^{-1}}$ and $\boxed{\tan^{-1}}$ functions on your calculator. They are the INVERSE functions of sin, cos and tan. You need to know their graphs, and their DOMAINS and RANGES.

**$y = \arcsin x$**

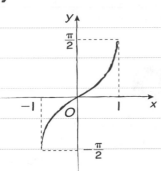

Domain: $-1 \leqslant x \leqslant 1$

Range: $-\dfrac{\pi}{2} \leqslant y \leqslant \dfrac{\pi}{2}$

**$y = \arccos x$**

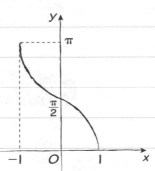

Domain: $-1 \leqslant x \leqslant 1$

Range: $0 \leqslant y \leqslant \pi$

**$y = \arctan x$**

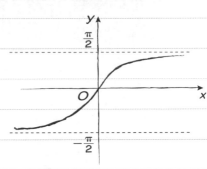

Domain: $-\infty < x < \infty$ (or $x \in \mathbb{R}$)

Range: $-\dfrac{\pi}{2} < y < \dfrac{\pi}{2}$

---

## Worked example

(a) Given that $y = \arcsin x$, express $\arccos x$ in terms of $y$. **(2 marks)**

$$x = \sin y$$
$$x = \cos\left(\frac{\pi}{2} - y\right)$$
$$\arccos x = \frac{\pi}{2} - y$$

(b) Hence evaluate $\arcsin x + \arccos x$. Give your answer in terms of $\pi$. **(2 marks)**

$$\arcsin x + \arccos x = y + \frac{\pi}{2} - y$$
$$= \frac{\pi}{2}$$

> You should always use **radians** when you are working with arcsin, arccos and arctan in your C3 exam.

## Worked example

The function f is defined by
$$f : x \mapsto \arctan x, \quad x \in \mathbb{R}$$

(a) State the range of f. **(1 mark)**

$$-\frac{\pi}{2} < f(x) < \frac{\pi}{2}$$

(b) Sketch the graph of $y = 2f(x) + \pi$. Show clearly any asymptotes and points of intersection with the coordinate axes. **(2 marks)**

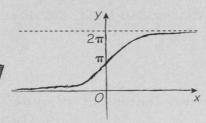

---

## Now try this

1. Write down the value of the following in radians.

   (a) $\arctan\sqrt{3}$

   (b) $\arccos\dfrac{1}{2}$

   (c) $\arcsin\dfrac{1}{\sqrt{2}}$ **(3 marks)**

> The domain of $g^{-1}$ is the range of g. Start with the range of arcsin $x$ then subtract $\dfrac{\pi}{4}$ from both limits. Remember to sketch the graph of $y = g^{-1}(x)$ only for values of $x$ in the domain of $g^{-1}$.

2. The function g is defined by
$$g : x \mapsto \arcsin x - \frac{\pi}{4}, \quad x \in \mathbb{R}, \quad -1 \leqslant x \leqslant 1$$

   (a) Find $g\left(\dfrac{1}{2}\right)$, giving your answer in terms of $\pi$. **(2 marks)**

   (b) Solve the equation $g(x) = 0$, giving your answer as an exact value. **(3 marks)**

   (c) Find $g^{-1}$ and state its domain. **(3 marks)**

   (d) Sketch the graph of $y = g^{-1}(x)$, showing the coordinates of the point where the graph crosses the $x$-axis. **(2 marks)**

# Addition formulae

These identities will appear in the C3 section of your formulae booklet :

$$\sin (A \pm B) = \sin A \cos B \pm \cos A \sin B$$

$$\cos (A \pm B) = \cos A \cos B \mp \sin A \sin B$$

$$\tan (A \pm B) = \frac{\tan A \pm \tan B}{1 \mp \tan A \tan B}$$

Each of these is actually TWO different identities. Be careful with $\pm$ and $\mp$. Take the top sign of each pair for one identity, and the bottom sign for the other.

$$\cos (A + B) = \cos A \cos B - \sin A \sin B$$
$$\cos (A - B) = \cos A \cos B + \sin A \sin B$$

---

$\cos x = \sin (90° - x)$ so $\cos 70° = \sin 20°$ and $\cos 20° = \sin 70°$. The answer contains $\tan 20°$, so write everything in terms of $\sin 20°$ and $\cos 20°$ before rearranging.

## Worked example

Given that $2\cos (x + 70°) = \sin (x + 20°)$, show, without using a calculator, that $\tan x = \frac{1}{3}\tan 20°$ **(4 marks)**

$2 (\cos x \cos 70° - \sin x \sin 70°)$
$\qquad = \sin x \cos 20° + \cos x \sin 20°$

$2\cos x \sin 20° - 2\sin x \cos 20°$
$\qquad = \sin x \cos 20° + \cos x \sin 20°$

$\cos x \sin 20° = 3 \sin x \cos 20°$

$\dfrac{\sin 20°}{\cos 20°} = \dfrac{3 \sin x}{\cos x}$

$\tan 20° = 3\tan x$ so $\tan x = \frac{1}{3}\tan 20°$

## Worked example

The diagram shows two right-angled triangles $ABC$ and $ADE$.

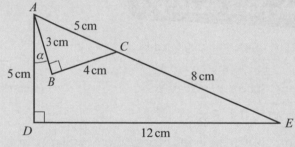

(a) Show that $\cos \alpha = \frac{63}{65}$ **(3 marks)**

$\cos \alpha = \cos (\angle EAD - \angle CAB)$

$\qquad = \cos \angle EAD \cos \angle CAB + \sin \angle EAD \sin \angle CAB$

$\qquad = \frac{5}{13} \times \frac{3}{5} + \frac{12}{13} \times \frac{4}{5}$

$\qquad = \frac{15}{65} + \frac{48}{65} = \frac{63}{65}$

(b) Hence, or otherwise, find the exact value of the length of $BD$. **(3 marks)**

$BD^2 = AB^2 + AD^2 - 2 \times AB \times AD \times \cos \alpha$

$\qquad = 3^2 + 5^2 - 2 \times 3 \times 5 \times \frac{63}{65}$

$\qquad = \frac{64}{13}$

$BD = \sqrt{\frac{64}{13}} = \frac{8}{\sqrt{13}}$

Be careful when applying the addition formulae for cos. The signs are **swapped** on the right-hand side.

## Hence or otherwise

If a question says 'hence or otherwise' it means you can use the earlier part of the question to help you find your answer. This is usually easier than starting this part of the question from scratch.

You might need to use techniques from C1 and C2 in your C3 exam. This question uses the formula for $\cos (A - B)$, then the **cosine rule** to find the missing length.

---

## Now try this

**1.** By writing $\sin 15° = \sin (45° - 30°)$,

show that $\operatorname{cosec} 15° = \dfrac{4}{\sqrt{6} - \sqrt{2}}$ **(4 marks)**

**2.** (a) Show that $\cos (x + 45°) = \dfrac{\cos x - \sin x}{\sqrt{2}}$ **(3 marks)**

(b) Hence solve, for $0 \leqslant x \leqslant 360°$,
$$\cos x - \sin x = 0.5$$
Give your answers in degrees to 1 decimal place. **(6 marks)**

# Double angle formulae

These double angle formulae can be DERIVED from the addition formulae. But they're really useful and you should LEARN them for your exam.

 **1**    $\sin 2A \equiv 2 \sin A \cos A$

 **2**    $\cos 2A \equiv \cos^2 A - \sin^2 A$
$\equiv 2 \cos^2 A - 1$
$\equiv 1 - 2 \sin^2 A$

**3**    $\tan 2A \equiv \dfrac{2 \tan A}{1 - \tan^2 A}$

You should learn all three versions of this one.

---

You need to be able to prove all of the identities above using the addition formulae on page 14. You will need to use the identity $\sin^2 A + \cos^2 A \equiv 1$ to get your expressions completely in terms of $\sin^2$ or $\cos^2$.

## Worked example

Use the identity
$$\cos(A + B) \equiv \cos A \cos B - \sin A \sin B$$
to show that
$$\cos 2A \equiv 1 - 2\sin^2 A \qquad \textbf{(2 marks)}$$

$\cos 2A \equiv \cos(A + A) \equiv \cos A \cos A - \sin A \sin A$
$\equiv \cos^2 A - \sin^2 A$
$\equiv (1 - \sin^2 A) - \sin^2 A$
$\equiv 1 - 2\sin^2 A$

---

## Worked example

(a) By writing $\sin 3\theta$ as $\sin(2\theta + \theta)$, show that
$\sin 3\theta = 3\sin\theta - 4\sin^3\theta$    **(5 marks)**

$\sin 3\theta = \sin(2\theta + \theta)$
$= \sin 2\theta \cos\theta + \cos 2\theta \sin\theta$
$= (2\sin\theta \cos\theta)\cos\theta + (1 - 2\sin^2\theta)\sin\theta$
$= 2\sin\theta \cos^2\theta + \sin\theta - 2\sin^3\theta$
$= 2\sin\theta(1 - \sin^2\theta) + \sin\theta - 2\sin^3\theta$
$= 2\sin\theta - 2\sin^3\theta + \sin\theta - 2\sin^3\theta$
$= 3\sin\theta - 4\sin^3\theta$

(b) Given that $\sin\theta = \dfrac{\sqrt{3}}{4}$, find the exact value of $\sin 3\theta$.    **(2 marks)**

$\sin 3\theta = 3\sin\theta - 4\sin^3\theta$
$= 3\left(\dfrac{\sqrt{3}}{4}\right) - 4\left(\dfrac{\sqrt{3}}{4}\right)^3$
$= \dfrac{3\sqrt{3}}{4} - \dfrac{3\sqrt{3}}{16}$
$= \dfrac{9\sqrt{3}}{16}$

---

As long as you aren't asked to prove a double angle formula in an exam question, it's OK to use it without deriving it. It's a good idea to show the steps when you use each identity clearly. You can use brackets to show the substitutions, or write down the identities you are using.

### Working backwards

Identities can be applied in BOTH DIRECTIONS. Make sure you are familiar enough with each of the double angle formulae to recognise EITHER SIDE when you see it.

Here are some useful variations:

• $\sin A \cos A = \frac{1}{2}\sin 2A$

• $\cos A + 1 = 2\cos^2 \frac{1}{2}A$

• $1 - \cos A = 2\sin^2 \frac{1}{2}A$

Part (b) says 'hence', so use the result from part (a) to get started. Look for similarities between the expression in part (a) and the equation in part (b).

---

 **Now try this**

1. Use the identity
$$\cos(A + B) \equiv \cos A \cos B - \sin A \sin B$$
to show that $\cos 2A = 2\cos^2 A - 1$    **(2 marks)**

2. (a) Show that $\cos 3x = 4\cos^3 x - 3\cos x$   **(5 marks)**
(b) Hence solve, for $0 \leqslant x \leqslant 90°$,
$$8\cos^3 x - 6\cos x - 1 = 0 \qquad \textbf{(3 marks)}$$

# $a\cos\theta \pm b\sin\theta$

You can use the addition formulae to write expressions of the form $a\cos\theta \pm b\sin\theta$ in the form $R\cos(\theta \mp \alpha)$ or $R\sin(\theta \pm \alpha)$. This can help you solve harder trig equations.

If you use these rules, be careful with the signs. You will usually be able to use POSITIVE values of $a$ and $b$.

## Golden rules

 $a\cos\theta \pm b\sin\theta = R\cos(\theta \mp \alpha)$

 $a\sin\theta \pm b\cos\theta = R\sin(\theta \pm \alpha)$

where $R = \sqrt{a^2 + b^2}$ and $\alpha = \arctan\left(\dfrac{b}{a}\right)$

## Worked example

(a) Express $5\cos x - 3\sin x$ in the form $R\cos(x + \alpha)$, where $R > 0$ and $0 < \alpha < \dfrac{\pi}{2}$. **(4 marks)**

$R = \sqrt{5^2 + 3^2} = \sqrt{34}$

$\alpha = \arctan\left(\dfrac{3}{5}\right) = 0.5404\ldots$

$5\cos x - 3\sin x = \sqrt{34}\cos(x + 0.5404)$

(b) Hence, or otherwise, solve the equation

$5\cos x - 3\sin x = 4$

for $0 \leqslant x \leqslant 2\pi$, giving your answers to 2 decimal places. **(5 marks)**

$\sqrt{34}\cos(x + 0.5404\ldots) = 4$

$\cos(x + 0.5404\ldots) = \dfrac{4}{\sqrt{34}}$

$x + 0.5404\ldots = 0.8148\ldots$

$x = 0.8148\ldots - 0.5404\ldots$

$= 0.2744\ldots$

or　$x + 0.5404\ldots = 2\pi - 0.8148\ldots$

$= 5.4683\ldots$

$x = 5.4683\ldots - 0.5404\ldots$

$= 4.9279\ldots$

$x = 0.27,\ 4.93$ (2 d.p.)

---

If you don't want to learn the rules above, you can use the addition formulae from page 14:

$5\cos x - 3\sin x = R\cos x \cos\alpha - R\sin x \sin\alpha$

Equate coefficients of $\cos x$:　$5 = R\cos\alpha$　①

Equate coefficients of $\sin x$:　$3 = R\sin\alpha$　②

①² + ②²: $R^2(\cos^2\alpha + \sin^2\alpha) = 5^2 + 3^2$

$R = \sqrt{5^2 + 3^2}$

② ÷ ①: $\dfrac{R\sin\alpha}{R\cos\alpha} = \tan\alpha = \dfrac{3}{5}$

$\alpha = \arctan\left(\dfrac{3}{5}\right)$

## EXAM ALERT!

Pay attention to these three Rs when solving trig equations in your C3 exam:

- **Rounding** – Don't round any values until the end of your calculation. Learn how to use the 'STORE' or 'MEMORY' functions on your calculator so you can use unrounded values, or write down values to at least 4 decimal places.
- **Radians** – Look at the range to decide whether you should be working in degrees or radians. Make sure your calculator is in the correct mode.
- **Range** – Check that all your solutions are within the specified range, and check that you have found every possible solution within that range.

Students have struggled with this topic in recent exams – be prepared!

Be careful. In part (b) of this question, the principal value from your calculator will not give you an answer in the required range.

## Now try this

1. (a) Express $3\sin 2\theta + 2\cos 2\theta$ in the form $R\sin(2\theta + \alpha)$, where $R > 0$ and $0 < \alpha < \dfrac{\pi}{2}$. **(4 marks)**

   (b) Hence, or otherwise, solve the equation

   $3\sin 2\theta + 2\cos 2\theta + 1 = 0$ for $0 \leqslant \theta \leqslant \pi$,

   giving your answers to 2 decimal places. **(5 marks)**

2. The function f is defined by:

   $y = \mathrm{f} : x \mapsto \sqrt{3}\cos x + \sin x$

   (a) Given that $\mathrm{f}(x) = R\cos(x - \alpha)$, where $R > 0$ and $0 < \alpha < 90°$, find the value of $R$ and the value of $\alpha$. **(4 marks)**

   (b) Hence sketch the graph of $y = \mathrm{f}(x)$ for $0 \leqslant x \leqslant 360°$, showing clearly the coordinates of any maxima or minima, and any points where the graph meets the coordinate axes. **(5 marks)**

# Trig modelling

Tides, circular motion, pendulums and springs are just a few examples of real-life situations that can be modelled using trigonometric functions.

## Worked example

(a) Express $2\sin\theta - 1.5\cos\theta$ in the form $R\sin(\theta - \alpha)$, where $R > 0$ and $0 < \alpha < \dfrac{\pi}{2}$. Give the value of $\alpha$ to 4 decimal places.

**(3 marks)**

$R = \sqrt{2^2 + 1.5^2} = 2.5$

$\alpha = \arctan\left(\dfrac{1.5}{2}\right) = 0.64350\ldots$

$2\sin\theta - 1.5\cos\theta = 2.5\sin(\theta - 0.6435)$

(b) (i) Find the maximum value of $2\sin\theta - 1.5\cos\theta$.

   (ii) Find the value of $\theta$, for $0 \leq \theta < \pi$, at which this maximum occurs. **(3 marks)**

(i)   2.5

(ii)   $\sin(\theta - 0.6435) = 1$

$$\theta - 0.6435 = \dfrac{\pi}{2}$$

$$\theta = 2.2143 \ (4 \text{ d.p.})$$

Tom models the height of sea water, $H$ metres, on a particular day by the equation

$$H = 6 + 2\sin\left(\dfrac{4\pi t}{25}\right) - 1.5\cos\left(\dfrac{4\pi t}{25}\right), \quad 0 \leq t < 12,$$

where $t$ is the number of hours after midday.

(c) Calculate the maximum value of $H$ predicted by this model and the value of $t$, to 2 decimal places, when this maximum occurs. **(3 marks)**

$H = 6 + 2.5\sin\left(\dfrac{4\pi t}{25} - 0.6435\right)$

$H_{\text{max}} = 8.5$

Occurs when $\dfrac{4\pi t}{25} = 2.2143$

so $t = 4.41$ (2 d.p.)

(d) Calculate, to the nearest minute, the times when the height of sea water is predicted, by this model, to be 7 metres. **(6 marks)**

$6 + 2.5\sin\left(\dfrac{4\pi t}{25} - 0.6435\right) = 7$

$\sin\left(\dfrac{4\pi t}{25} - 0.6435\right) = 0.4$

$\dfrac{4\pi t}{25} - 0.6435 = 0.4115\ldots \text{ or } \pi - 0.4115\ldots$

$t = 2.0988\ldots$      $t = 6.7115\ldots$

   $= 2.06\,\text{pm}$       $= 6.43\,\text{pm}$

## Maximum and minimum

A function in the form $R\cos(\theta \pm \alpha)$ or $R\sin(\theta \pm \alpha)$ has a maximum value of $R$ and a minimum value of $-R$.

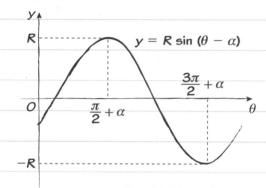

The maximum value of $R\sin(\theta - \alpha)$ occurs when $\sin(\theta - \alpha) = 1$.

This occurs when $\theta - \alpha = \dfrac{\pi}{2}, \dfrac{5\pi}{2}, \ldots$

There will often be a connection between different parts of the same exam question. Look for similarities between the complicated equation in part (b) and the expression in part (a).

Part (d) says 'times' so make sure you find all possible solutions for the range of the model. Remember to convert to minutes. Multiply the decimal part by 60:
$0.0988\ldots \times 60 = 5.9\ldots$

## Now try this

The displacement, $d$ cm, of a pendulum at time $t$ seconds is modelled by the equation

$$d = 4\cos 1.2t + 2\sin 1.2t, \quad t > 0$$

(a) Given that $d = R\cos(1.2t - \alpha)$, where $R > 0$ and $0 < \alpha < \dfrac{\pi}{2}$, find the value of $R$ and the value of $\alpha$. **(3 marks)**

(b) Write down the maximum displacement of the pendulum. **(3 marks)**

(c) Find the times in the interval $0 < t < 5$ when the displacement of the pendulum is 0. **(2 marks)**

# Exponential functions

You need to know about these two functions, and to be able to sketch their graphs.

 **1** $f(x) = e^x, \quad x \in \mathbb{R}$

This is called the EXPONENTIAL FUNCTION. The letter 'e' represents a constant number, and is equal to 2.71828...

 **2** $g(x) = \ln x, \quad x \in \mathbb{R}, \quad x > 0$

This is called the NATURAL LOGARITHM. It is a logarithm to the base e. You could write $g(x) = \log_e x$. It is the INVERSE of the exponential function.

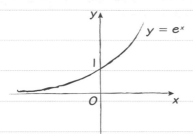
$y = e^x$

Asymptote at $y = 0$ ✓     Crosses y-axis at (0, 1) ✓

$y = \ln x$

Asymptote at $x = 0$ ✓     Crosses x-axis at (1, 0) ✓

## Worked example

The functions f and g are defined by
$f : x \mapsto 1 + e^{2x}, \quad x \in \mathbb{R}$
$g : x \mapsto \ln x, \quad x \in \mathbb{R}, \quad x > 0$

(a) Find fg and state its range. **(3 marks)**

$fg(x) = 1 + e^{2(\ln x)}$
$\qquad = 1 + e^{\ln(x^2)}$
$\qquad = 1 + x^2$
$x^2 > 0$ so range of fg is $fg(x) > 1$

(b) Find $f^{-1}$, stating its domain. **(4 marks)**

$y = 1 + e^{2x}$
$e^{2x} = y - 1$
$2x = \ln(y - 1)$
$x = \frac{1}{2}\ln(y - 1)$

$f^{-1}(x) = \frac{1}{2}\ln(x - 1)$
$e^{2x} > 0$ so range of f is $f(x) > 1$
So domain of $f^{-1}$ is $x > 1$

(c) Sketch on the same axes the curves with equations $y = f(x)$ and $y = f^{-1}(x)$, giving the coordinates of all points where the curves cross the coordinate axes. **(4 marks)**

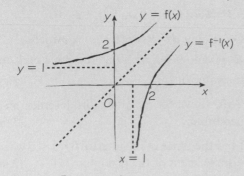

## Golden rule

$\ln x$ works just like any other logarithm:
$e^{\ln x} = \ln(e^x) = x$

Look for the $e^{\blacksquare}$ and $\boxed{\ln}$ functions on your calculator to find values of $e^x$ and $\ln x$.

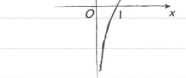 You need to use the laws of logs: $2 \ln x = \ln x^2$

For part (b) write the function in the form $y = ...$ then rearrange to make x the subject. You can rearrange an equation involving an exponential by getting the exponential on its own on one side, then **taking logs** of both sides. Using the golden rule above, $\ln(e^{2x}) = 2x$. Use brackets to show you are taking the logs of **everything** on the right-hand side.

## Now try this

1. Given that $f(x) = \ln x$, $x > 0$, sketch, on separate axes, the graphs of
   (a) $y = |f(x)|$ **(2 marks)**
   (b) $y = f(x - 2)$ **(2 marks)**
   (c) $y = -f(2x)$ **(2 marks)**

   In each case show the point where the graph meets or crosses the x-axis and state the equation of the asymptote.

2. The point P with y-coordinate 6 lies on the curve with equation $y = 3e^{2x - 1}$. Find, in terms of $\ln 2$, the x-coordinate of P. **(2 marks)**

Start with $6 = 3e^{2x-1}$. Divide both sides by 3 then take the natural logarithm of both sides.

# Exponential equations

You need to be able to use the laws of indices and logs to manipulate and solve equations involving $e^x$ and $\ln x$. Make sure you are confident with solving C2 log equations before revising this topic.

## Worked example

Find the exact solutions to the equations

(a) $\ln x + \ln 3 = \ln 6$      **(2 marks)**

$\ln 3x = \ln 6$

  $3x = 6$

    $x = 2$

(b) $e^x + 3e^{-x} = 4$      **(4 marks)**

    $(e^x)^2 + 3 = 4e^x$

$(e^x)^2 - 4e^x + 3 = 0$

$(e^x - 3)(e^x - 1) = 0$

$e^x = 3$        $e^x = 1$

$x = \ln 3$     $x = \ln 1 = 0$

## EXAM ALERT!

If a question asks for **exact solutions** then you should leave your answers as logs, or powers of e. Don't write your answers as decimals. Make sure you simplify your logs and powers as much as possible. Remember:

$$e^0 = 1 \qquad \ln 1 = 0$$

In part (b), you can convert to a quadratic using the fact that $e^{-x} \times e^x = 1$. You can also sometimes use $e^{2x} = (e^x)^2$ to write a quadratic in $e^x$.

> Students have struggled with this topic in recent exams – be prepared!

## Laws of logs

Here is a reminder of the most useful laws of logarithms for your C3 exam:

**1**   $\ln a + \ln b = \ln ab$

**2**   $\ln a - \ln b = \ln \dfrac{a}{b}$

**3**   $\ln \left(\dfrac{1}{a}\right) = -\ln a$

**4**   $a \ln b = \ln (b^a)$

Check whether both answers are within the range given in the question. The equation is only valid for $-1 < x < 2$ so $x = 5$ is not a valid solution.

You can't combine the $2x$ with the $e^{3x+1}$ easily, so just take logs of both sides. You can then use the laws of logs to simplify the left-hand side. Group the $x$ terms together, then factorise to get $x$ on its own.

## Worked example

Find algebraically the exact solutions to the equations

(a) $\ln (4 - 2x) + \ln (9 - 3x) = 2\ln (x + 1), \; -1 < x < 2$      **(5 marks)**

$\ln [(4 - 2x)(9 - 3x)] = \ln (x + 1)^2$

  $(4 - 2x)(9 - 3x) = (x + 1)^2$

  $6x^2 - 30x + 36 = x^2 + 2x + 1$

    $5x^2 - 32x + 35 = 0$

     $(5x - 7)(x - 5) = 0$

       $x = \dfrac{7}{5}$      $x = 5$ not valid

(b) $2^x e^{3x+1} = 10$

Give your answer to part (b) in the form $\dfrac{a + \ln b}{c + \ln d}$

where $a$, $b$, $c$ and $d$ are integers.      **(5 marks)**

    $\ln[2^x e^{3x+1}] = \ln 10$

  $\ln 2^x + \ln e^{3x+1} = \ln 10$

    $x \ln 2 + 3x + 1 = \ln 10$

      $x(\ln 2 + 3) = \ln 10 - 1$

         $x = \dfrac{-1 + \ln 10}{3 + \ln 2}$

## Now try this

1. Solve

  (a) $\ln (x + 1) - \ln x = \ln 5$      **(2 marks)**

  (b) $e^{4x} + 3e^{2x} = 10$      **(5 marks)**

  (c) $\ln (6x + 7) = 2 \ln x, \; x > 0$      **(4 marks)**

2. Find the exact solution to the equation

    $3^x e^{2x-5} = 7$      **(5 marks)**

3. The function f is defined by

$$f : x \mapsto \frac{3x^2 - 7x + 2}{x^2 - 4}, \quad x \neq \pm 2$$

  (a) Show that $f(x) = \dfrac{3x - 1}{x + 2}$      **(3 marks)**

  (b) Hence, or otherwise, solve the equation

    $\ln (3x^2 - 7x + 2) = 1 + \ln (x^2 - 4), \; x > 2$

    giving your answer in terms of e.      **(4 marks)**

# Exponential modelling

You can use exponential functions to model lots of real-life situations.

**1** **Growth models**

A typical growth model can be described as $N = C + N_0 e^{kt}$

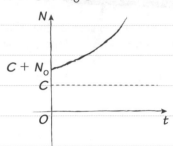

**2** **Decay models**

A typical decay model can be described as $N = C + N_0 e^{-kt}$

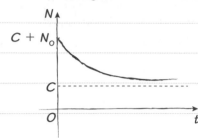

---

## Worked example

This model describes the temperature $P\,°C$ of the water in a kettle $t$ minutes after it has boiled:
$$P = 25 + A\,e^{-kt}, \quad t \geq 0$$
where $A$ and $k$ are positive constants.

(a) Given that the initial temperature of the water was $100\,°C$, find the value of $A$.    **(2 marks)**

When $t = 0$, $P = 25 + Ae^0$

$\quad 100 = 25 + A$

$\quad\quad A = 75$

After 10 minutes, the water in the kettle has cooled down to $40\,°C$.

(b) Show that $k = \frac{1}{10}\ln 5$.    **(3 marks)**

$\quad 40 = 25 + 75\,e^{-10k}$

$\quad 15 = 75\,e^{-10k}$

$\quad \frac{1}{5} = e^{-10k}$

$\quad -10k = \ln\left(\frac{1}{5}\right) = \ln(5^{-1}) = -\ln 5$

So $k = \frac{1}{10}\ln 5$

(c) Find the temperature of the water after 18 minutes, in $°C$ to 1 decimal place.    **(2 marks)**

$P = 25 + 75\,e^{-\left(\frac{1}{10}\ln 5\right)\times 18} = 29.1\,°C$ (1 d.p.)

You will sometimes be given a complete exponential model and asked to use it. In this example, you are not given two of the constants in the model. To find them:

- Substitute the information given in the question into the equation of the model. Make sure you substitute the right variable in the right place.
- Solve an equation to find the unknown constants.

### Rates of change

You might have to find the RATE OF CHANGE of a quantity in an exponential model. You can do this by DIFFERENTIATING with respect to time. In the model in the Worked example on the left, the rate of change of temperature with time would be given by
$$\frac{dP}{dt} = -kA\,e^{-kt}$$
The units would be $°C$/min.

Revise differentiation like this on pages 21 and 22.

---

## Now try this

1. The number of cells, $N$, in a bacterial culture at a time $t$ hours after midday is modelled as
$$N = 4000 + 100\,e^{0.8t}, \quad t \geq 0$$

   (a) Write down the number of cells in the culture at midday.    **(1 mark)**

   (b) Find the time at which the culture contains 8000 cells. Give your answer to the nearest minute.    **(5 marks)**

2. The mass, $M$ grams, of a sample of radon after $t$ hours is modelled using the equation
$$M = 250\,e^{-kt}, \quad t \geq 0$$
where $k$ is a positive constant.

   (a) What was the initial mass of the sample?    **(1 mark)**

   After 90 hours the sample has lost half its mass.

   (b) Find the value of $k$ to 3 significant figures.    **(4 marks)**

# The chain rule

The chain rule lets you differentiate a FUNCTION OF A FUNCTION. The chain rule is not in the formulae booklet, so make sure you know how to use it confidently. Follow this step-by-step method:

 Substitue the 'inside' function with $u$.

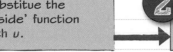 Treat $u$ as a single variable and differentiate.

 Multiply the result by the derivative of $u$.

$$y = \sqrt{x^3 + 1}$$
$$u = x^3 + 1$$
$$y = \sqrt{u} = u^{\frac{1}{2}}$$

$$\frac{dy}{du} = \frac{1}{2}u^{-\frac{1}{2}}$$
$$= \frac{1}{2}(x^3 + 1)^{-\frac{1}{2}}$$
$$= \frac{1}{2\sqrt{x^3 + 1}}$$

$$\frac{du}{dx} = 3x^2$$
$$\frac{dy}{dx} = \frac{dy}{du} \times \frac{du}{dx} = \frac{3x^2}{2\sqrt{x^3 + 1}}$$

This is how the chain rule is sometimes stated.

## Worked example

$f(x) = \dfrac{2}{3x + 1}$,  $x > 0$

Differentiate $f(x)$ and find $f'(1)$.  **(3 marks)**

$y = f(x) = 2(3x + 1)^{-1}$

$\quad = 2u^{-1}$ with $u = 3x + 1$

$\dfrac{dy}{dx} = \dfrac{dy}{du} \times \dfrac{du}{dx} = -2u^{-2} \times 3$

$\quad = -6(3x + 1)^{-2}$

$f'(x) = \dfrac{-6}{(3x + 1)^2}$ $\qquad f'(1) = \dfrac{-6}{(3 \times 1 + 1)^2} = -\dfrac{3}{8}$

You don't have to write down all of this working to apply the chain rule in your exam. If you're confident you can jump straight from the first line of working to:

$f'(x) = -2(3x + 1)^{-2} \times 3$

## Splodge

You can use the splodge method to apply the chain rule quickly. The 'inside' function is splodge. You work out $\dfrac{dy}{d(\text{splodge})}$, then multiply by the derivative of splodge!

$$y = ((5x - 1))^9$$

$$\frac{dy}{dx} = 5 \times 9((5x - 1))^8$$

Derivative of splodge

$\dfrac{dy}{d(\text{splodge})}$

## Worked example

$C$ is a curve with equation $2y^3 + 10y + 2 = x$

(a) Find $\dfrac{dy}{dx}$ in terms of $y$.  **(3 marks)**

$\dfrac{dx}{dy} = 6y^2 + 10$ so $\dfrac{dy}{dx} = \dfrac{1}{6y^2 + 10}$

(b) Find the gradient of the curve at the point $(-8, -2)$.  **(2 marks)**

$y = -2$, so $\dfrac{dy}{dx} = \dfrac{1}{6(-2)^2 + 10} = \dfrac{1}{34}$

## Functions of $y$

You can use this version of the chain rule to differentiate equations where $x$ is given in terms of $y$: $\dfrac{dy}{dx} = \dfrac{1}{\left(\dfrac{dx}{dy}\right)}$

$\dfrac{dy}{dx}$ is given in terms of $y$ so you need to substitute the $y$-coordinate to find the gradient.

## Now try this

1. Given that $y = \dfrac{1}{\sqrt{x^2 - 3x + 1}}$, find $\dfrac{dy}{dx}$.  **(3 marks)**

2. $f(x) = (\sqrt[3]{x} + 6)^6$. Differentiate $f(x)$ and find $f'(8)$, writing your answer as a power of 2.  **(4 marks)**

# Derivatives to learn

You need to LEARN these four derivatives for your C3 exam. They're not given in the formulae booklet and you need to be really confident using them.

## Trigonometric functions

**1**   $\dfrac{d}{dx}(\sin x) = \cos x$

**2**   $\dfrac{d}{dx}(\cos x) = -\sin x$

These rules ONLY work for angles measured in RADIANS.

## Exponential functions

**3**   $\dfrac{d}{dx}(e^x) = e^x$

**4**   $\dfrac{d}{dx}(\ln x) = \dfrac{1}{x}$

$e^x$ is the ONLY function that is the same when differentiated.

### Worked example

Differentiate with respect to $x$

(a) $\sin(x^2 + 1)$     (2 marks)

$\dfrac{d}{dx}[\sin(x^2 + 1)] = 2x\cos(x^2 + 1)$

(b) $\sec x$     (3 marks)

$$\frac{d}{dx}(\sec x) = \frac{d}{dx}(\cos x)^{-1}$$
$$= -(\cos x)^{-2} \times (-\sin x)$$
$$= \frac{\sin x}{\cos^2 x} = \sec x \tan x$$

You need the chain rule for both of these.

### Round in circles

If you keep differentiating $y = \sin x$ you will end up back where you started.

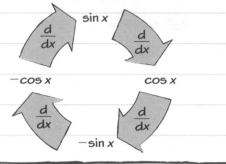

### Worked example

The function f is defined by
$$f : x \mapsto \ln(4 - x^2), \quad 0 \leq x < 2$$
Find the exact value of the gradient of the curve $y = f(x)$ at the point where it crosses the $x$-axis.

(5 marks)

$f'(x) = \dfrac{1}{4 - x^2} \times (-2x) = \dfrac{2x}{x^2 - 4}$

When $f(x) = 0$, $\ln(4 - x^2) = 0$

$$4 - x^2 = 1$$
$$x = \sqrt{3} \text{ or } x = \cancel{-\sqrt{3}}$$

$f'(\sqrt{3}) = \dfrac{2\sqrt{3}}{(\sqrt{3})^2 - 4} = -2\sqrt{3}$

When you differentiate a log function in your C3 exam, the derivative will usually not contain any logs or exponentials.

### Worked example

The value of a boat, £$V$, after $t$ years is modelled as
$$V = 12\,000\,e^{-\frac{1}{5}t}, \quad t > 0$$

(a) Find $\dfrac{dV}{dt}$     (2 marks)

$$\frac{dV}{dt} = 12\,000\,e^{-\frac{1}{5}t} \times \left(-\frac{1}{5}\right)$$
$$= -2400\,e^{-\frac{1}{5}t}$$

(b) Find the exact value of $V$ when $\dfrac{dV}{dt} = -800$.     (3 marks)

$$-800 = -2400\,e^{-\frac{1}{5}t}$$
$$e^{-\frac{1}{5}t} = \frac{1}{3}$$
$$-\frac{1}{5}t = \ln\left(\frac{1}{3}\right) = \ln 3$$
$$t = 5\ln 3$$

You will need to use $\dfrac{dy}{dx} = \dfrac{1}{\left(\frac{dx}{dy}\right)}$, and then use

$\sin^2 2y + \cos^2 2y = 1$ to write the right-hand side in terms of $x$.

### Now try this

1. Differentiate with respect to $x$

   (a) $\sin 2x - \cos 4x$     (2 marks)

   (b) $x^2 - e^{4x-3}$     (3 marks)

   (c) $\ln(3x^2 + 1)$     (2 marks)

2. Given that $x = \sin 2y$, show that
$$\frac{dy}{dx} = \frac{1}{2\sqrt{1 - x^2}}$$
    (4 marks)

3. Use the chain rule to show that $\dfrac{d}{dx}(\sin^2 x) = \sin 2x$
    (3 marks)

# The product rule

The product rule lets you differentiate two functions MULTIPLIED together. It's not given in the formulae booklet so you need to LEARN it. Here's an example of a function that can be differentiated using the product rule:

$$y = \underbrace{e^x}_{u}\underbrace{(x^2 - 2)}_{v}$$

**Golden rule**

If $y = uv$, where $u$ and $v$ are functions of $x$, then

$$\frac{dy}{dx} = u\frac{dv}{dx} + v\frac{du}{dx}$$

In function notation, if $h(x) = f(x)\,g(x)$, then
$h'(x) = f(x)\,g'(x) + g(x)\,f'(x)$

## Worked example

Differentiate with respect to $x$

(a) $y = x^4 \ln 3x$                    **(3 marks)**

$u = x^4$          $v = \ln 3x$

$\frac{du}{dx} = 4x^3$      $\frac{dv}{dx} = \frac{1}{x}$

$\frac{dy}{dx} = u\frac{dv}{dx} + v\frac{du}{dx} = x^4\left(\frac{1}{x}\right) + (\ln 3x)(4x^3)$

$\qquad = x^3(1 + 4\ln 3x)$

(b) $y = e^x(1 + \cos 2x)$              **(3 marks)**

$u = e^x$          $v = 1 + \cos 2x$

$\frac{du}{dx} = e^x$      $\frac{dv}{dx} = -2\sin 2x$

$\frac{dy}{dx} = u\frac{dv}{dx} + v\frac{du}{dx} = e^x(-2\sin 2x) + (1 + \cos 2x)\,e^x$

$\qquad = e^x(1 - 2\sin 2x + \cos 2x)$

For any constant, $k$
$\frac{d}{dx}(\ln kx) = \frac{1}{kx} \times k = \frac{1}{x}$

## EXAM ALERT!

It's easy to get in a mess with the product rule.

Start by writing out $u$, $v$, $\frac{du}{dx}$ and $\frac{dv}{dx}$.

Then use brackets when you substitute to make sure you don't make a mistake.

Students have struggled with this topic in recent exams – be prepared!

## Formulae booklet

Unless you are specifically asked to show them, you can quote these results from the formulae booklet:

| f(x) | f'(x) |
|------|-------|
| $\tan kx$ | $k\sec^2 kx$ |
| $\sec x$ | $\sec x \tan x$ |
| $\cot x$ | $-\csc^2 x$ |
| $\csc x$ | $-\csc x \cot x$ |

For part (b) you need to look at the factors of $h'(x)$. $e^{2x}$ and $\csc x$ can never equal 0, so the factor $(2 - \cot x)$ must equal 0.

## Worked example

(a) Given that $h(x) = e^{2x}\csc x$, find $h'(x)$.
                                        **(4 marks)**

$h(x) = f(x)\,g(x)$

$f(x) = e^{2x}$          $g(x) = \csc x$

$f'(x) = 2e^{2x}$        $g'(x) = -\csc x \cot x$

$h'(x) = f(x)\,g'(x) + g(x)\,f'(x)$

$\qquad = e^{2x}(-\csc x \cot x) + (\csc x)(2e^{2x})$

$\qquad = e^{2x}\csc x(2 - \cot x)$

(b) Solve, in the interval $0 \leq x < \frac{\pi}{2}$, the equation $h'(x) = 0$.          **(4 marks)**

$e^{2x}\csc x(2 - \cot x) = 0$

$e^{2x} \neq 0 \qquad \csc x \neq 0 \qquad 2 - \cot x = 0$

$\qquad\qquad\qquad\qquad\qquad \tan x = \frac{1}{2}$

$\qquad\qquad\qquad\qquad\qquad x = 0.464 \text{ (3 s.f.)}$

## Now try this

1. Differentiate with respect to $x$

   (a) $\sqrt{x}\sin 5x$              **(3 marks)**

   (b) $\ln x^x$                      **(3 marks)**

 Write $\ln x^x$ as $x\ln x$

2. A curve has equation $y = e^x(3x^2 - 4x - 1)$.

   (a) Find $\frac{dy}{dx}$.          **(4 marks)**

   (b) Find the $x$-coordinates of the turning points on the curve.          **(4 marks)**

# The quotient rule

If one function is DIVIDED by another you can differentiate them using the quotient rule. One version is given in the formulae booklet:

$$\underset{\text{function}}{\frac{f(x)}{g(x)}} \qquad \underset{\text{derivative}}{\frac{f'(x)\,g(x) - f(x)\,g'(x)}{(g(x))^2}}$$

But the version on the right is often easier to use, so you should LEARN it for your exam.

### Golden rule

If $y = \dfrac{u}{v}$, where $u$ and $v$ are functions of $x$, then

$$\frac{dy}{dx} = \frac{v\dfrac{du}{dx} - u\dfrac{dv}{dx}}{v^2}$$

The minus sign on top of the fraction means the ORDER is important.

## Worked example

Differentiate $\dfrac{\sin 5x}{x^2}$ with respect to $x$.    **(3 marks)**

$u = \sin 5x$ $\qquad\qquad$ $v = x^2$

$\dfrac{du}{dx} = 5\cos 5x$ $\qquad$ $\dfrac{dv}{dx} = 2x$

$\dfrac{dy}{dx} = \dfrac{v\dfrac{du}{dx} - u\dfrac{dv}{dx}}{v^2} = \dfrac{(x^2)(5\cos 5x) - (\sin 5x)(2x)}{(x^2)^2}$

$\qquad = \dfrac{5x\cos 5x - 2\sin 5x}{x^3}$

When using the quotient rule, start by writing out $u$, $v$, $\dfrac{du}{dx}$ and $\dfrac{dv}{dx}$. Then use brackets when you substitute to make sure you don't make a mistake.

Make sure you are really confident differentiating expressions like $\sin 5x$ using the chain rule.

This solution shows the quotient rule using function notation. You will have to answer questions about curves or graphs in the form $y = \ldots$ **and** questions about functions in the form $f : x \mapsto \ldots$ so it's a good idea to be comfortable with both versions.

## Worked example

The function h is defined by

$$h : x \mapsto \frac{e^x + 2}{e^x - 3}, \quad x \in \mathbb{R}, \quad x \neq \ln 3$$

Show that $h'(x) = \dfrac{-5e^x}{(e^x - 3)^2}$    **(3 marks)**

$h(x) = \dfrac{f(x)}{g(x)}$

$f(x) = e^x + 2$ $\qquad\qquad$ $g(x) = e^x - 3$

$f'(x) = e^x$ $\qquad\qquad\qquad$ $g'(x) = e^x$

$h'(x) = \dfrac{f'(x)\,g(x) - f(x)\,g'(x)}{[g(x)]^2}$

$\qquad = \dfrac{e^x(e^x - 3) - (e^x + 2)\,e^x}{(e^x - 3)^2}$

$\qquad = \dfrac{-5e^x}{(e^x - 3)^2}$

Be really careful with signs when using the quotient rule. The minus sign on top means you have to subtract **all** of $f(x)\,g'(x)$.

## Now try this

1. Differentiate with respect to $x$

   (a) $\dfrac{\cos x}{\sqrt{x}}$    **(3 marks)**

   (b) $\dfrac{2x}{\sqrt{3x + 1}}$    **(3 marks)**

   (c) $\dfrac{x^2 + 1}{\ln(x^2 + 1)}$    **(4 marks)**

2. Use the quotient rule to show that

   $\dfrac{d}{dx}(\cot x) = -\text{cosec}^2 x$    **(4 marks)**

3. $f(x) = \dfrac{3}{x - 1} - \dfrac{3}{x^2 + 2} - \dfrac{9}{(x^2 + 2)(x - 1)}$

   (a) Show that $f(x) = \dfrac{3x}{x^2 + 2}$    **(4 marks)**

   (b) Hence, or otherwise, find $f'(x)$ in its simplest form.    **(3 marks)**

# Differentiation and graphs

In your C3 exam you might have to use $\dfrac{dy}{dx}$ to find the gradient of a curve at a given point. You can use this information to find turning points, and to find the equations of tangents and normals to the curve.

## Worked example

A curve $C$ has equation
$$y = \frac{3}{(5-3x)^2}, \quad x \neq \frac{5}{3}$$

The point $P$ on $C$ has $x$-coordinate 2. Find an equation of the normal to $C$ at $P$ in the form $ax + by + c = 0$, where $a$, $b$ and $c$ are integers.

**(7 marks)**

$y = 3(5-3x)^{-2}$

$\dfrac{dy}{dx} = -6(5-3x)^{-3} \times (-3)$

$= 18(5-3x)^{-3}$

$= \dfrac{18}{(5-3x)^3}$

At $P$, $\dfrac{dy}{dx} = \dfrac{18}{(5-3(2))^3} = -18$

At $P$, $y = \dfrac{3}{(5-3(2))^2} = 3$

so $P$ is the point $(2, 3)$

Gradient of normal at $P = \dfrac{-1}{-18} = \dfrac{1}{18}$

$y - 3 = \dfrac{1}{18}(x - 2)$

$18y - 54 = x - 2$

$x - 18y + 52 = 0$

## Tangents and normals

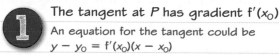

If the point $P(x_0, y_0)$ lies on the curve with equation $y = f(x)$:

**1** The tangent at $P$ has gradient $f'(x_0)$

An equation for the tangent could be
$$y - y_0 = f'(x_0)(x - x_0)$$

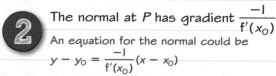

**2** The normal at $P$ has gradient $\dfrac{-1}{f'(x_0)}$

An equation for the normal could be
$$y - y_0 = \frac{-1}{f'(x_0)}(x - x_0)$$

## EXAM ALERT!

It's easier to use the **chain rule** than the **quotient rule** to differentiate a function like this. Rewrite it with a negative power.

Make sure you read the question carefully. You need to find the **normal** and not the **tangent**, and you will lose a mark if you don't give your final equation in the correct form.

Students have struggled with this topic in recent exams – be prepared!

---

The curve $y = f(x)$ has **turning points** when $\dfrac{dy}{dx} = 0$. To find the coordinates of $P$, you need to solve the equation $e^{2x}(1 + \tan x)^2 = 0$.

## Now try this

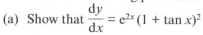

The diagram shows a sketch of the curve with equation $y = e^{2x} \tan x$, $-\dfrac{\pi}{2} < x < \dfrac{\pi}{2}$

The curve has a turning point at $P$.

(a) Show that $\dfrac{dy}{dx} = e^{2x}(1 + \tan x)^2$ **(3 marks)**

(b) Find the exact coordinates of $P$. **(3 marks)**

(c) Find an equation of the tangent to the curve at the point where $x = 1$. Give your answer in the form $y = ax + b$ where $a$ and $b$ are constants given to 3 significant figures. **(3 marks)**

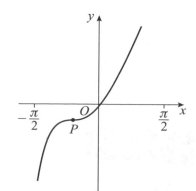

# Iteration

An iterative process is one where an answer is fed back in as a new starting value. You can use an ITERATIVE FORMULA to find numerical answers to a given degree of accuracy. You will be told if you have to use iteration in your C3 exam, and you will be given a formula.

## Worked example

$f(x) = x^3 + 2x^2 - 3x - 11$

(a) Show that $f(x) = 0$ can be rearranged as

$$x = \sqrt{\frac{3x + 11}{x + 2}}, \quad x \neq -2 \qquad \textbf{(2 marks)}$$

$x^3 + 2x^2 - 3x - 11 = 0$

$x^2(x + 2) - 3x - 11 = 0$

$x^2(x + 2) = 3x + 11$

$x^2 = \dfrac{3x + 11}{x + 2}$ so $x = \sqrt{\dfrac{3x + 11}{x + 2}}$

The equation $f(x) = 0$ has one positive root, $\alpha$.

The iterative formula $x_{n+1} = \sqrt{\dfrac{3x_n + 11}{x_n + 2}}$ is used to find an approximation to $\alpha$.

(b) Taking $x_1 = 0$, find, to 3 decimal places, the values of $x_2$, $x_3$ and $x_4$.    **(3 marks)**

$x_2 = \sqrt{\dfrac{3(0) + 11}{(0) + 2}} = 2.34520788... = 2.345 \ (3 \ \text{d.p.})$

$x_3 = \sqrt{\dfrac{3x_2 + 11}{x_2 + 2}} = 2.03732494... = 2.037 \ (3 \ \text{d.p.})$

$x_4 = \sqrt{\dfrac{3x_3 + 11}{x_3 + 2}} = 2.05874811... = 2.059 \ (3 \ \text{d.p.})$

(c) Show that $\alpha = 2.057$ correct to 3 decimal places.    **(3 marks)**

$f(2.0565) = (2.0565)^3 + 2(2.0565)^2 - 3(2.0565) - 11$

$\qquad = -0.01378...$ Negative

$f(2.0575) = (2.0575)^3 + 2(2.0575)^2 - 3(2.0575) - 11$

$\qquad = 0.0041401...$ Positive

Change of sign so $2.0565 < \alpha < 2.0575$, so $\alpha = 2.057$ correct to 3 d.p.

## Change of sign

You can use a change of sign (from POSITIVE to NEGATIVE, or vice versa) to show that a particular interval contains a root of an equation.

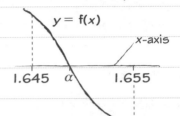

$f(1.645)$ is positive and $f(1.655)$ is negative so $f(x) = 0$ has a root, $\alpha$, between 1.645 and 1.655. All values in this interval round to 1.65, so $\alpha = 1.65$ to 2 decimal places.

You might have to rearrange an equation before using iteration. The final form has a square root sign, so take a factor of $x^2$ out of the first two terms of the polynomial.

## EXAM ALERT!

When you use a change of sign to find a root, remember to use the **original function**. You **have** to write a conclusion, so make sure you write 'change of sign' and write down the interval that contains the root.

> Students have struggled with this topic in recent exams – be prepared!

## Using a calculator

You can find iteration values quickly on your calculator. Once you've worked out the first value, replace $x_n$ with the [ Ans ] function and just keep pressing [ = ]

$$\sqrt{\frac{3\text{Ans} + 11}{\text{Ans} + 2}}$$

$$2.03732494$$

## Now try this

$f(x) = \ln(x + 1) - 2x + 2, \quad x > 0$

The equation $f(x) = 0$ has one root, $\alpha$.

(a) Show that $1 < \alpha < 2$.    **(2 marks)**

(b) Use the iterative formula

$$x_{n+1} = \tfrac{1}{2}\ln(x_n + 1) + 1, \quad x_0 = 1.5$$

to calculate values of $x_1$, $x_2$ and $x_3$, giving your answers to 5 decimal places.    **(3 marks)**

(c) Show that $\alpha = 1.4475$ correct to 4 decimal places.    **(3 marks)**

# You are the examiner!

CHECKING YOUR WORK is one of the key skills you will need for your C3 exam. All five of these students have made ONE key mistake in their working. Can you spot them all?

**1** Express $\dfrac{3}{2x+3} - \dfrac{1}{2x-3} + \dfrac{6}{4x^2-9}$ as a single fraction in its simplest form. **(4 marks)**

$$\dfrac{3}{2x+3} - \dfrac{1}{2x-3} + \dfrac{6}{(2x+3)(2x-3)}$$
$$= \dfrac{3(2x-3) - 2x + 3 + 6}{(2x+3)(2x-3)}$$
$$= \dfrac{4x}{(2x+3)(2x-3)}$$

**2** Given that
$$3\cos\theta - 6\sin\theta = R\cos(\theta + \alpha),$$
where $R > 0$ and $0 < \alpha < 90°$,
find the exact value of $R$ and the value of $\alpha$ correct to 2 decimal places. **(3 marks)**

$R = \sqrt{3^2 + 6^2} = 6.71$ (2 d.p.)

$\alpha = \arctan\left(\dfrac{6}{3}\right) = 63.43°$ (2 d.p.)

**3** Given that $f(x) = 3e^x + 2$, $x \in \mathbb{R}$, sketch the curve with equation $y = f(x)$.

Show any points at which the curve cuts or meets the axes and the equations of any asymptotes. **(3 marks)**

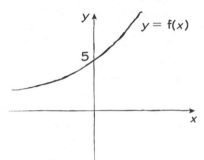

**4** Given that
$$y = \dfrac{x^2 - 1}{e^{3x}}, \quad x > 1, \text{ find } \dfrac{dy}{dx}. \quad \textbf{(3 marks)}$$

$$\dfrac{dy}{dx} = \dfrac{(x^2 - 1)(3e^{3x}) - (e^{3x})(2x)}{(e^{3x})^2}$$
$$= \dfrac{3x^2 - 2x - 3}{e^{3x}}$$

**5** Solve the equation
$$\ln(2x-1) + \ln(x-1) = 2\ln(x+1), \quad x > 1 \quad \textbf{(5 marks)}$$

$$\ln(2x-1) + \ln(x-1) - \ln(x+1)^2 = 0$$
$$\ln\dfrac{(2x-1)(x-1)}{(x+1)^2} = 0$$
$$\dfrac{(2x-1)(x-1)}{(x+1)^2} = 0$$

So $x = 1$ or $x = \dfrac{1}{2}$

## Checking your work

If you have any time left at the end of your exam, you should check back through your working.

☑ Check you have answered EVERY PART and given all the information asked for.

☑ Double check any ADDITION FORMULAE to make sure you have + and − in the right places.

☑ Make sure everything is EASY TO READ.

☑ Cross out any incorrect working with a SINGLE NEAT LINE and UNDERLINE the correct answer.

☑ Make sure any sketches are LABELLED.

**Now try this**

Find the mistake in each student's answer on this page, and write out the correct working for each question. Turn over for the answers.

# You are still the examiner!

BEFORE looking at this page, turn back to page 27 and try to spot the key mistake in each student's working. Use this page to CHECK your answers — the corrections are shown in red, and these answers are now 100% CORRECT.

 **1**

Express $\dfrac{3}{2x+3} - \dfrac{1}{2x-3} + \dfrac{6}{4x^2-9}$ as a single fraction in its simplest form.

**(4 marks)**

$$\dfrac{3}{2x+3} - \dfrac{1}{2x-3} + \dfrac{6}{(2x+3)(2x-3)}$$

$$= \dfrac{3(2x-3) - (2x+3) + 6}{(2x+3)(2x-3)}$$

$$= \dfrac{4x}{(2x+3)(2x-3)}$$

$$= \dfrac{2(2x-3)}{(2x+3)(2x-3)}$$

$$= \dfrac{2}{2x+3}$$

**Top tip**

Always use **brackets** when working with algebraic fractions.

Revise this topic on page 1.

 **2**

Given that
$3\cos\theta - 6\sin\theta = R\cos(\theta + \alpha)$,
where $R > 0$ and $0 < \alpha < 90°$,
find the exact value of $R$ and the value of $\alpha$ correct to 2 decimal places.

**(3 marks)**

$$R = \sqrt{3^2 + 6^2} = \cancel{6.71\ (2\ d.p.)} = 3\sqrt{5}$$

$$\alpha = \arctan\left(\dfrac{6}{3}\right) = 63.43°\ (2\ d.p.)$$

**Top tip**

Read the question carefully. If it asks for an exact value then leave your answer as a surd.

Have a look at page 16.

**Top tip**

Always write out the product rule and the quotient rule before substituting. It also helps to write down your $u$, $v$, $\dfrac{du}{dx}$ and $\dfrac{dv}{dx}$.

Revise the quotient rule on page 24.

 **3**

Given that $f(x) = 3e^x + 2$, $x \in \mathbb{R}$, sketch the curve with equation $y = f(x)$. Show any points at which the curve cuts or meets the axes and the equations of any asymptotes.

**(3 marks)**

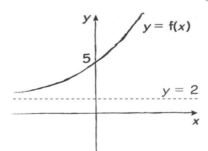

**4**

Given that
$$y = \dfrac{x^2 - 1}{e^{3x}}, \quad x > 1, \text{ find } \dfrac{dy}{dx}.$$

**(3 marks)**

$$u = x^2 - 1 \qquad\qquad v = e^{3x}$$
$$\dfrac{du}{dx} = 2x \qquad\qquad \dfrac{dv}{dx} = 3e^{3x}$$

$$\dfrac{dy}{dx} = \dfrac{v\dfrac{du}{dx} - u\dfrac{dv}{dx}}{v^2} = \dfrac{(x^2-1)(3e^{3x}) - (e^{3x})(2x)}{(e^{3x})^2}$$

$$= \dfrac{3x^2 - 2x - 3}{e^{3x}} = \dfrac{-3x^2 + 2x + 3}{e^{3x}}$$

**Top tip**

Make sure you **fully label** any sketches of graphs. You need to draw **asymptotes** and give their equations.

Graphs of exponential functions are on page 18.

**5**

Solve the equation
$\ln(2x-1) + \ln(x-1) = 2\ln(x+1), \quad x > 1$

**(5 marks)**

$$\ln(2x-1) + \ln(x-1) - \ln(x+1)^2 = 0$$

$$\ln\dfrac{(2x-1)(x-1)}{(x+1)^2} = 0$$

$$\dfrac{(2x-1)(x-1)}{(x+1)^2} \cancel{= 0} = 1$$

 ~~So $x = 1$ or $x = \frac{1}{2}$~~

$$(2x-1)(x-1) = (x+1)^2$$

$$2x^2 - 3x + 1 = x^2 + 2x + 1$$

$$x^2 - 5x = 0$$

$$x(x-5) = 0$$

 So ~~$x = 0$ or~~ $x = 5$

**Top tip**

Be careful with logs. You have to rearrange then undo the logs **on both sides**. The right-hand side is 1 because $e^0 = 1$. Remember to dismiss any solutions that are not in the range $x > 1$.

Logs and exponential equations are covered on page 19.

# Partial fractions

Many algebraic fractions can be written as the SUM of simpler fractions. This technique is called writing a fraction in PARTIAL FRACTIONS. In your C4 exam you will probably have to use partial fractions to write a SERIES EXPANSION or to INTEGRATE. You can revise these topics on pages 32 and 38.

## Worked example

$f(x) = \dfrac{7 - 2x}{(2x - 1)(x + 1)}$

Express f(x) in partial fractions.   **(3 marks)**

$\dfrac{7 - 2x}{(2x - 1)(x + 1)} = \dfrac{A}{2x - 1} + \dfrac{B}{x + 1}$

$7 - 2x = A(x + 1) + B(2x - 1)$

Let $x = \frac{1}{2}$:    $7 - 2\left(\frac{1}{2}\right) = A\left(\frac{1}{2} + 1\right)$

$\underline{A = 4}$

Let $x = -1$:    $7 - 2(-1) = B(2(-1) - 1)$

$\underline{B = -3}$

$f(x) = \dfrac{4}{2x - 1} - \dfrac{3}{x + 1}$

## Golden rule

Find as many missing values as possible by SUBSTITUTING values for $x$ to make some of the factors equal to zero. The more factors you can find this way, the easier it will be to EQUATE COEFFICIENTS later.

The denominators on the right-hand side are factors of the original denominator. If all the factors are different, then each one appears as a denominator once.

## Cover up and calculate

If the expression has NO REPEATED FACTORS, you can use this quick method to find numerators. Choose a factor, and work out the value of $x$ which makes that factor equal to zero. Then cover it up with your finger, and evaluate what's left of the fraction with that value of $x$.

$f(x) = \dfrac{7 - 2x}{(2x - 1)(\phantom{x+1})}$    $\dfrac{7 - 2(-1)}{2(-1) - 1} = \dfrac{9}{-3} = -3$

Covering up $(x + 1)$ in the Worked example above and evaluating what's left with $x = -1$ gives you $B$.

## EXAM ALERT!

You need to do a bit more work if there is a repeated factor, like $(x - 3)^2$, or if the fraction is improper. You should always work out any values you can by substituting first. Here you can work out one value by substituting $x = 3$. To work out the other values you need to **equate coefficients**. You could multiply out both sides first:

$2x^2 - 1 = Ax^2 + (-6A + B)x + (9A - 3B + C)$

Students have struggled with this topic in recent exams – be prepared!   ✎

## Worked example

$\dfrac{2x^2 - 1}{(x - 3)^2} = A + \dfrac{B}{(x - 3)} + \dfrac{C}{(x - 3)^2}, \quad x \neq 3$

Find the values of $A$, $B$ and $C$.   **(4 marks)**

$2x^2 - 1 = A(x - 3)^2 + B(x - 3) + C$

Let $x = 3$:    $2(3)^2 - 1 = C$

$\underline{C = 17}$

Equate $x^2$ terms:    $\underline{A = 2}$

Equate constant terms: $-1 = 9A - 3B + C$

$-1 = 18 - 3B + 17$

$-36 = -3B$

$\underline{B = 12}$

## Now try this

There is a repeated factor, so you need one fraction with denominator $(x + 2)$ and another fraction with denominator $(x + 2)^2$.

**1.** $g(x) = \dfrac{8x^2}{(3x - 2)(x + 2)^2}$

Express g(x) in partial fractions.   **(4 marks)**

**2.** $\dfrac{6x^2 - 1}{(2x - 3)(x + 1)} = A + \dfrac{B}{(2x - 3)} + \dfrac{C}{(x + 1)}$

Find the values of $A$, $B$ and $C$.   **(4 marks)**

# Parametric equations

You can define a curve by giving the $x$-coordinate and $y$-coordinate as separate functions of another variable, called a PARAMETER.

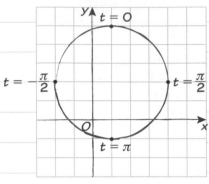

## Golden rules

**1** Each value of $t$ generates a point on the curve.

**2** The value of $t$ is NEITHER the $x$- NOR $y$-coordinate of the curve.

The curve with parametric equations
$$y = 3\cos t + 2, \quad x = 3\sin t + 1, \quad -\pi < t \leqslant \pi$$
is a circle, centre $(1, 2)$ with radius 3.

---

## Converting to cartesian

Cartesian equations only involve $x$ and $y$. To convert to cartesian form you need to ELIMINATE the parameter, $t$.

**1** If the equations contain trig functions use an identity like $\sin^2 t + \cos^2 t = 1$.

**2** Otherwise, write $t$ in terms of either $x$ or $y$ then substitute into the other equation.

The curve crosses the $y$-axis when $x = 0$. Substitute $x = 0$ into the $x$-equation to find the value of $t$ at this point. Then substitute this value of $t$ into the $y$-equation to find the value of $y$ at this point.

Find an expression for $\cos t$ in terms of $x$ and an expression for $\sin t$ in terms of $y$. You can then use $\sin^2 t + \cos^2 t = 1$ to eliminate $t$. Finally, rearrange the cartesian equation into the form asked for in the question.

### Worked example

The curve $C$ has parametric equations
$$x = \ln t, \quad y = t^2 - 3$$
Find a cartesian equation of $C$.   **(3 marks)**

$t = e^x$

$y = (e^x)^2 - 3 = e^{2x} - 3$

Start by using the addition formula for $\sin(A + B)$ to write $y$ in terms of $\sin t$ and $\cos t$.

### Worked example

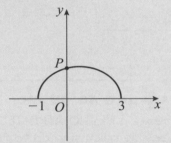

The curve $C$ shown has parametric equations
$$x = 1 + 2\cos t, \quad y = \sin t, \quad 0 \leqslant t \leqslant \pi$$

(a) Find the exact coordinates of the point $P$ where the curve crosses the $y$-axis.   **(3 marks)**

When $x = 0$, $1 + 2\cos t = 0$
$$\cos t = -\frac{1}{2}$$
$$t = \frac{2\pi}{3}$$

When $t = \frac{2\pi}{3}$, $y = \sin\frac{2\pi}{3} = \frac{\sqrt{3}}{2}$

$P$ is $\left(0, \frac{\sqrt{3}}{2}\right)$

(b) Find a cartesian equation of $C$ in the form $y^2 = f(x)$.   **(3 marks)**

$$\cos t = \frac{x-1}{2} \qquad \sin t = y$$
$$\cos^2 t + \sin^2 t = 1$$
$$\left(\frac{x-1}{2}\right)^2 + y^2 = 1$$
$$y^2 = 1 - \left(\frac{x-1}{2}\right)^2$$

### Now try this

$$x = \sin t, \quad y = \sin\left(t + \frac{\pi}{3}\right), \quad -\frac{\pi}{2} < t < \frac{\pi}{2}$$

Show that a cartesian equation of the curve is
$$y = \frac{1}{2}x + \frac{\sqrt{3}}{2}\sqrt{1 - x^2}, \quad -1 < x < 1 \qquad \textbf{(3 marks)}$$

# Parametric differentiation

You can use this version of the CHAIN RULE to find the gradient of a curve defined by PARAMETRIC EQUATIONS.

This formula gives you the derivative IN TERMS OF $t$. You need to know the value of $t$ for a particular point on the curve to find the gradient of the curve at that point.

**Golden rule**

$$\frac{dy}{dx} = \frac{dy}{dt} \div \frac{dx}{dt}$$

## Worked example

A curve $C$ has parametric equations

$$x = \sin^2 t, \; y = 2\tan t, \quad 0 \le t < \frac{\pi}{2}$$

(a) Find $\dfrac{dy}{dx}$ in terms of $t$.    **(4 marks)**

$$\frac{dx}{dt} = 2\sin t \cos t, \; \frac{dy}{dt} = 2\sec^2 t$$

$$\frac{dy}{dx} = \frac{2\sec^2 t}{2\sin t \cos t} = \frac{1}{\sin t \cos^3 t}$$

The tangent to $C$ at the point where $t = \frac{\pi}{3}$ cuts the $x$-axis at the point $P$.

(b) Find the $x$-coordinate of $P$.    **(6 marks)**

At the point where $t = \frac{\pi}{3}$:

$$x = \sin^2\left(\frac{\pi}{3}\right) = \left(\frac{\sqrt{3}}{2}\right)^2 = \frac{3}{4}$$

$$y = 2\tan\left(\frac{\pi}{3}\right) = 2\sqrt{3}$$

$$\frac{dy}{dx} = \frac{1}{\sin\left(\frac{\pi}{3}\right)\cos^3\left(\frac{\pi}{3}\right)} = \frac{1}{\left(\frac{\sqrt{3}}{2}\right)\left(\frac{1}{8}\right)} = \frac{16}{\sqrt{3}}$$

Equation of tangent:

$$y - 2\sqrt{3} = \frac{16}{\sqrt{3}}\left(x - \frac{3}{4}\right)$$

At $P$, $y = 0$, so $-2\sqrt{3} = \frac{16}{\sqrt{3}}\left(x - \frac{3}{4}\right)$

$$-\frac{3}{8} = x - \frac{3}{4}$$

$$x = \frac{3}{8}$$

If you have to differentiate a curve given in parametric form you will usually be asked to find $\dfrac{dy}{dx}$ in terms of $t$. Find $\dfrac{dx}{dt}$ and $\dfrac{dy}{dt}$ and write them down. Then work out $\dfrac{dy}{dt} \div \dfrac{dx}{dt}$. You don't have to simplify your answer, but it might speed up your working later in the question.

Here is a sketch of the curve $C$ and the tangent at the point where $t = \frac{\pi}{3}$:

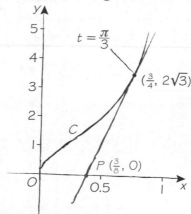

## Now try this

The diagram shows a sketch of the curve $C$ with parametric equations

$$x = 2\cos\left(t - \frac{\pi}{3}\right), \; y = 3\sin 2t, \qquad 0 \le t < 2\pi$$

(a) Find $\dfrac{dy}{dx}$ in terms of $t$.    **(4 marks)**

(b) Find an equation of the normal to the curve at the point where $t = 0$.    **(4 marks)**

(c) Find the coordinates of all the points on $C$ where $\dfrac{dy}{dx} = 0$.    **(5 marks)**

Find all the values of $t$ where $\dfrac{dy}{dx} = 0$, then find the $x$- and $y$- coordinates at each of these points.

# The binomial series

You need to be able to use the binomial theorem to find a SERIES EXPANSION of expressions in the form $(a + bx)^n$, where $n$ is ANY REAL NUMBER. You need to use this version of the binomial series, which is given in the C2 section of your formulae booklet:

$$(1 + x)^n = 1 + nx + \frac{n(n-1)}{1 \times 2}x^2 + \dots + \frac{n(n-1)\dots(n-r+1)}{1 \times 2 \times \dots \times r}x^r + \dots \qquad (|x| < 1, n \in \mathbb{R})$$

If you are given an expression in the form $(a + bx)^n$ you will need to rearrange it by taking

out a factor of $a^n$ like this: $(a + bx)^n = a^n\left(1 + \frac{bx}{a}\right)^n$

## Worked example

$f(x) = \dfrac{1}{\sqrt{4 + 5x}}, \quad |x| < \dfrac{4}{5}$

(a) Find the binomial expansion of $f(x)$, in ascending powers of $x$, as far as the term in $x^3$, giving each coefficient as a simplified fraction. **(6 marks)**

$f(x) = (4 + 5x)^{-\frac{1}{2}} = 4^{-\frac{1}{2}}\left(1 + \dfrac{5x}{4}\right)^{-\frac{1}{2}}$

$= \dfrac{1}{2}\left(1 + \dfrac{5x}{4}\right)^{-\frac{1}{2}}$

$= \dfrac{1}{2}\left[1 + \left(-\dfrac{1}{2}\right)\left(\dfrac{5x}{4}\right) + \dfrac{\left(-\frac{1}{2}\right)\left(-\frac{3}{2}\right)}{1 \times 2}\left(\dfrac{5x}{4}\right)^2 + \dfrac{\left(-\frac{1}{2}\right)\left(-\frac{3}{2}\right)\left(-\frac{5}{2}\right)}{1 \times 2 \times 3}\left(\dfrac{5x}{4}\right)^3 + \dots\right]$

$= \dfrac{1}{2}\left[1 - \dfrac{5}{8}x + \dfrac{75}{128}x^2 - \dfrac{625}{1024}x^3 + \dots\right]$

$= \dfrac{1}{2} - \dfrac{5}{16}x + \dfrac{75}{256}x^2 - \dfrac{625}{2048}x^3 + \dots$

(b) Hence find the coefficient of $x$ in the series expansion of $\dfrac{3 + x}{\sqrt{4 + 5x}}$ **(4 marks)**

$\dfrac{3 + x}{\sqrt{4 + 5x}} = (3 + x)\left(\dfrac{1}{2} - \dfrac{5}{16}x + \dfrac{75}{256}x^2 - \dfrac{625}{2048}x^3 + \dots\right)$

$x$ term in expansion $= (3)\left(-\dfrac{5}{16}x\right) + (x)\left(\dfrac{1}{2}\right)$

$= \left(-\dfrac{15}{16} + \dfrac{1}{2}\right)x = -\dfrac{7}{16}x$

So coefficient of $x$ is $-\dfrac{7}{16}$

> Start by writing the expression in the form $(a + bx)^n$, then take out a factor of $a^n$. You can now use the binomial series from the formulae booklet, with $n = -\dfrac{1}{2}$, replacing $x$ with $\dfrac{5x}{4}$.

> A common mistake is not to square or cube **all** of $\dfrac{5x}{4}$, so use brackets when you write out the series. You can work out the coefficients in one go on your calculator using brackets and the 💾 key.

## Now try this

(a) Expand $\sqrt[3]{1 - 6x}$, $\quad |x| < \dfrac{1}{3}$

in ascending powers of $x$ up to and including the $x^3$ term, simplifying each term. **(4 marks)**

(b) Use your expansion, with a suitable value of $x$, to obtain an approximation to $\sqrt[3]{0.94}$. Give your answer to 6 decimal places. **(2 marks)**

> You must use your answer to part (a). To find a suitable value of $x$, solve $1 - 6x = 0.94$.

# Implicit differentiation

An IMPLICIT equation is one that cannot be easily written in the form $y = f(x)$ or $x = g(y)$.
You can differentiate an implicit equation to find an expression for $\dfrac{dy}{dx}$ in terms of $x$ and $y$. Follow these steps:

- Differentiate EVERY TERM on BOTH SIDES of the equation WITH RESPECT TO $x$.
- Collect terms involving $\dfrac{dy}{dx}$ on one side and the remaining terms on the other.
- Factorise to get $\dfrac{dy}{dx}$ on its own.

### Golden rules

These rules will help you with implicit differentiation in your C4 exam:

**1** $\dfrac{d}{dx}[f(y)] = f'(y)\dfrac{dy}{dx}$

$\dfrac{d}{dx}[e^{2y}] = 2e^{2y}\dfrac{dy}{dx}$

**2** $\dfrac{d}{dx}[g(x)y] = g'(x)y + g(x)\dfrac{dy}{dx}$

$\dfrac{d}{dx}[x^3 y] = 3x^2 y + x^3\dfrac{dy}{dx}$

## EXAM ALERT!

Make sure you differentiate the **constant term** on the right-hand side:

$$\dfrac{d}{dx}[1] = 0$$

You can use golden rule 1 above to differentiate $3\cos 2y$ with respect to $x$. Remember that you have $x$ **and** $y$ in your expression for $\dfrac{dy}{dx}$ so you probably won't be able to simplify it very much.

> Students have struggled with this topic in recent exams – be prepared!

## Worked example

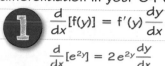

A set of curves satisfy the equation

$$6\sin 2x + 3\cos 2y = 1$$

Find $\dfrac{dy}{dx}$ in terms of $x$ and $y$. **(3 marks)**

$$12\cos 2x - 6\sin 2y\,\dfrac{dy}{dx} = 0$$

$$\dfrac{dy}{dx} = \dfrac{2\cos 2x}{\sin 2y}$$

Be careful with the $6xy$ term. You can use the product rule (golden rule 2 above). Once you have differentiated every term, you need to rearrange the equation so all the $\dfrac{dy}{dx}$ terms are on one side. You can then factorise to get $\dfrac{dy}{dx}$ on its own, then divide by the other factor. To find the gradient at $(1, -2)$ you need to substitute $x = 1$ and $y = -2$ into your expression for $\dfrac{dy}{dx}$.

## Worked example

A curve $C$ is described by the equation

$$3x^2 + 4y^2 - 2x + 6xy - 5 = 0$$

Find an equation of the tangent to $C$ at the point $(1, -2)$, giving your answer in the form $ax + by + c = 0$, where $a$, $b$ and $c$ are integers. **(7 marks)**

$$6x + 8y\,\dfrac{dy}{dx} - 2 + 6y + 6x\,\dfrac{dy}{dx} = 0$$

$$8y\,\dfrac{dy}{dx} + 6x\,\dfrac{dy}{dx} = 2 - 6y - 6x$$

$$\dfrac{dy}{dx}(8y + 6x) = 2 - 6y - 6x$$

$$\dfrac{dy}{dx} = \dfrac{2 - 6y - 6x}{8y + 6x}$$

$$= \dfrac{1 - 3y - 3x}{4y + 3x}$$

At the point $(1, -2)$, $\dfrac{dy}{dx} = \dfrac{1 - 3(-2) - 3(1)}{4(-2) + 3(1)} = -\dfrac{4}{5}$

Equation of tangent: $y - (-2) = -\dfrac{4}{5}(x - 1)$

$$4x + 5y + 6 = 0$$

## Now try this

1. The point $P$ with coordinates $(3, -1)$ lies on the curve with equation $x^3 + y^2 + 3x^2 y = 1$.

   Show that at $P$, $\dfrac{dy}{dx} = -\dfrac{9}{25}$ **(5 marks)**

2. The curve $C$ is described by the equation
   $$2x^2 - y^2 = ye^{3x}$$

   (a) Show that the point $(0, -1)$ lies on $C$. **(1 mark)**

   (b) Find an equation of the tangent to $C$ at the point $(0, -1)$. **(7 marks)**

$$\dfrac{d}{dx}[ye^{3x}] = 3ye^{3x} + e^{3x}\dfrac{dy}{dx}$$

# Differentiating $a^x$

In your C3 exam you used $\frac{d}{dx}[e^x] = e^x$ to differentiate expressions involving $e^x$. You need to be more careful when differentiating powers of constants OTHER THAN $e$.

$$\frac{d}{dx}[a^x] = a^x \ln a$$

You can learn this rule but you need to know how to DERIVE it as well. Follow the steps in the box on the right to use $e^x$ to differentiate $a^x$.

## Two methods

**1** You can use the laws of logs to differentiate $a^x$.

$y = a^x = e^{x \ln a}$

$\frac{dy}{dx} = \ln a \, e^{x \ln a} = a^x \ln a$

**2** You can differentiate $a^x$ implicitly.

$y = a^x$

$\ln y = \ln a^x = x \ln a$

$\frac{1}{y}\frac{dy}{dx} = \ln a$

$\frac{dy}{dx} = y \ln a = a^x \ln a$

## Worked example

Given that $y = 3^x$, show that $\frac{dy}{dx} = 3^x \ln 3$   **(2 marks)**

$y = e^{x \ln 3}$

$\frac{dy}{dx} = \ln 3 \, e^{x \ln 3}$

$\quad = 3^x \ln 3$

Make sure you're confident converting between $a^x$ and $e^{x \ln a}$:

$a^x = e^{\ln a^x} = e^{x \ln a}$

## Worked example

A curve has equation $2^x + y^2 = 2xy$.

Find the exact value of $\frac{dy}{dx}$ at the point on $C$ with coordinates $(3, 2)$.   **(7 marks)**

$2^x \ln 2 + 2y\frac{dy}{dx} = 2y + 2x\frac{dy}{dx}$

$2y\frac{dy}{dx} - 2x\frac{dy}{dx} = 2y - 2^x \ln 2$

$\frac{dy}{dx}(2y - 2x) = 2y - 2^x \ln 2$

$\frac{dy}{dx} = \frac{2y - 2^x \ln 2}{2y - 2x}$

At the point $(3, 2)$, $\frac{dy}{dx} = \frac{2(2) - 2^3 \ln 2}{2(2) - 2(3)}$

$\quad = -2 + 4 \ln 2$

You might have to differentiate $a^x$ as part of an implicit differentiation. As long as you're not asked to show it, you can use the rule:

$\frac{d}{dx}[a^x] = a^x \ln a$

Be really careful when you're substituting with implicit differentiation. Make sure you substitute the $x$- and $y$-values in the correct places in your expression for $\frac{dy}{dx}$.

There's more on implicit differentiation on page 33.

## Now try this

1. Find $\frac{d}{dx}[4^x \sin x]$   **(3 marks)**

Use the product rule.
Look at page 23 for a reminder.

2. A curve has equation $xy + \left(\frac{1}{2}\right)^y = 2$.

Find an equation of the normal to the curve at the point $(0, -1)$.   **(6 marks)**

The gradient of the normal will be $\dfrac{-1}{\left(\frac{dy}{dx}\right)}$

# Rates of change

You can model lots of physical or financial situations by describing how a variable changes with time. Equations involving rates of change are called DIFFERENTIAL EQUATIONS. You can revise how to FORM differential equations on this page, and how to SOLVE them on page 46.

## Worked example

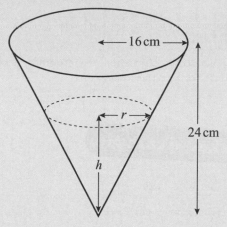

A container is made in the shape of a hollow inverted right circular cone. The height of the container is 24 cm and the radius is 16 cm, as shown in the diagram. Water is flowing into the container. When the height of water is $h$ cm, the surface of the water has radius $r$ cm and the volume of water is $V$ cm³.

(a) Show that $V = \dfrac{4\pi h^3}{27}$.    **(2 marks)**

[*The volume $V$ of a right circular cone with vertical height $h$ and base radius $r$ is given by the formula $V = \frac{1}{3}\pi r^2 h$.*]

$$\frac{r}{h} = \frac{16}{24} \text{ so } r = \frac{2h}{3}$$

$$V = \tfrac{1}{3}\pi r^2 h = \tfrac{1}{3}\pi\left(\frac{2h}{3}\right)^2 h$$

$$= \frac{4\pi h^3}{27}$$

Water flows into the container at a rate of 8 cm³ s⁻¹.

(b) Find, in terms of $\pi$, the rate of change of $h$ when $h = 12$.    **(5 marks)**

$$\frac{dV}{dh} = \frac{12\pi h^2}{27} = \frac{4\pi h^2}{9}$$

$$\frac{dh}{dt} = \frac{dV}{dt} \div \frac{dV}{dh} = 8 \div \frac{4\pi h^2}{9}$$

$$= 8 \times \frac{9}{4\pi h^2}$$

$$= \frac{18}{\pi h^2}$$

When $h = 12$:

$$\frac{dh}{dt} = \frac{18}{\pi (12)^2} = \frac{1}{8\pi}$$

If you need to use the formula for the volume of a cone in your exam it will be given to you with the question. You can use similar triangles to find the relationship between $r$ and $h$. Draw a sketch to help you.

### Chain rule

The chain rule allows you to multiply and divide derivatives in the same way as fractions.

$$\frac{dV}{dt} = \frac{dV}{dh} \times \frac{dh}{dt} \quad \text{so} \quad \frac{dh}{dt} = \frac{dV}{dt} \div \frac{dV}{dh}$$

Use your answer to part (a) to find $\dfrac{dV}{dh}$.

The rate the water is flowing into the container is $\dfrac{dV}{dt} = 8 \text{ cm}^3\text{s}^{-1}$. The rate of change of $h$ is $\dfrac{dh}{dt}$.

For part (a), remember that $\theta$ is a constant.

## Now try this

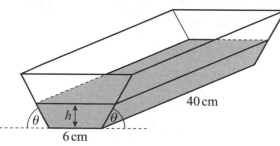

The diagram shows a section of gutter in the shape of a prism. The cross-section of the gutter is a symmetrical trapezium. Water is flowing into the gutter. When the depth of the water is $h$ cm, the volume of the water is $V$ cm³.

(a) Show that $\dfrac{dV}{dh} = 240 + 80h \cot\theta$    **(3 marks)**

Water flows into the gutter at a constant rate of 40 cm³ s⁻¹.

(b) Given that when $h = 2.5$ the rate of change of $h$ is 0.1 cm s⁻¹, find the value of $\theta$ correct to 1 decimal place.    **(5 marks)**

# Integrals to learn

Make sure you are really familiar with these four integrals. They're not given in the formulae booklet and you will use them a lot in your C4 exam.

## Trigonometric functions

**1** $\int \cos x \, dx = \sin x + c$

**2** $\int \sin x \, dx = -\cos x + c$

Remember that indefinite integrations always include a CONSTANT OF INTEGRATION, $c$.

## Exponential functions

**3** $\int e^x dx = e^x + c$

**4** $\int \left(\frac{1}{x}\right) dx = \ln|x| + c$

You use a modulus sign when you integrate $\frac{1}{x}$, because $\ln x$ only exists for positive values of $x$.

---

### Integrating f(ax + b)

This is one of the most useful integration rules for your C4 exam:

$$\int f'(ax + b) \, dx = \frac{1}{a} f(ax + b) + c$$

This is an application of the chain rule 'in reverse'. There are more examples of this on the next page.

**Worked example**

(a) Find $\int \cos 2x \, dx$   **(2 marks)**

$\int \cos 2x \, dx = \frac{1}{2} \sin 2x + c$

(b) Find $\int \frac{6}{3x + 1} dx$   **(2 marks)**

$\int \frac{6}{3x + 1} dx = 2 \ln|3x + 1| + c$

$$\int \frac{6}{3x + 1} dx = 6 \int \frac{1}{3x + 1} dx$$

---

This is a **definite integration**. You don't need to use a constant of integration.

**Worked example**

Show that $\int_0^1 \sqrt{2 - x} \, dx = \frac{4\sqrt{2} - 2}{3}$   **(3 marks)**

$\int_0^1 \sqrt{2 - x} \, dx = \int_0^1 (2 - x)^{\frac{1}{2}} dx$

$= \left[-\frac{2}{3}(2 - x)^{\frac{3}{2}}\right]_0^1$

$= \left(-\frac{2}{3}(1)^{\frac{3}{2}}\right) - \left(-\frac{2}{3}(2)^{\frac{3}{2}}\right)$

$= -\frac{2}{3} + \frac{2}{3}(2\sqrt{2}) = \frac{4\sqrt{2} - 2}{3}$

**EXAM ALERT!**

You can use the rule for integrating f(ax + b) to integrate any of these functions:

$f(x) = \frac{1}{ax + b}$     $f(x) = (ax + b)^n$

$f(x) = \sin(ax + b)$     $f(x) = \cos(ax + b)$

Remember to multiply by $\frac{1}{a}$ and leave $ax + b$ inside the function. You can always check your answers by differentiating.

Students have struggled with this topic in recent exams – be prepared!

---

### Now try this

1. (a) Find $\int 4 \sin 2x \, dx$   **(2 marks)**
   (b) Find $\int e^{\frac{x}{6}} dx$   **(2 marks)**

2. Find the exact value of $\int_0^{\frac{\pi}{2}} 2 \sin \frac{1}{2} \theta \, d\theta$   **(3 marks)**

3. Find $\int \tan 3x \, dx$   **(2 marks)**

Use this result from the formulae booklet:

| f(x) | $\int$f(x) d$x$ |
| --- | --- |
| $\tan x$ | $\ln|\sec x|$ |

# Reverse chain rule

You can use the chain rule IN REVERSE to integrate some expressions. Here are two useful examples. If you can spot these integrations you can save time in your exam:

 $\displaystyle\int f'(x)[f(x)^n]\,dx = \frac{1}{n+1}[f(x)]^{n+1} + c$    **2** $\displaystyle\int \frac{f'(x)}{f(x)}\,dx = \ln|f(x)| + c$

For example:   $n = 10$

$$\int 2x(x^2+1)^{10}\,dx = \frac{1}{11}(x^2+1)^{11} + c$$

$f'(x) = 2x \qquad f(x) = x^2 + 1$

For example:   $f'(x) = 6x^2$

$$\int \frac{6x^2}{2x^3-1}\,dx = \ln|2x^3-1| + c$$

$f(x) = 2x^3 - 1$

---

## Worked example

(a) Find $\displaystyle\int \frac{x}{5x^2+1}\,dx$     **(2 marks)**

$$\int \frac{x}{5x^2+1}\,dx = \frac{1}{10}\int \frac{10x}{5x^2+1}\,dx$$

$$= \frac{1}{10}\ln|5x^2+1| + c$$

(b) Find $\displaystyle\int x^2\sqrt{1-4x^3}\,dx$     **(3 marks)**

$$\int x^2\sqrt{1-4x^3}\,dx = \int x^2(1-4x^3)^{\frac{1}{2}}\,dx$$

Try $y = (1-4x^3)^{\frac{3}{2}}$

$$\frac{dy}{dx} = -12x^2 \times \frac{3}{2}(1-4x^3)^{\frac{1}{2}} = -18x^2(1-4x^3)^{\frac{1}{2}}$$

So $\displaystyle\int x^2\sqrt{1-4x^3}\,dx = -\frac{1}{18}(1-4x^3)^{\frac{3}{2}} + c$

## Adjusting constants

If you can guess the FORM of the integral, but can't work out what constant to multiply it by, then you can:

**1** Differentiate your guess for the integral.

**2** Compare it with the original expression.

**3** Adjust the constant if necessary.

For part (b) it's hard to see what constant to take outside the integral. Instead, write a guess for the integral as '$y = \dots$'

Now find $\dfrac{dy}{dx}$ and compare it with $x^2(1-4x^3)^{\frac{1}{2}}$.

It is $-18$ times the original integrand, so **divide** $y$ by $-18$.

---

## EXAM ALERT!

Watch out for common standard derivatives, especially when trigonometric functions are involved. You might need to use the results from the C3 section of the formulae booklet to spot the form of the integral.

| $f(x)$ | $f'(x)$ |
|--------|---------|
| $\sec x$ | $\sec x \tan x$ |

Students have struggled with this topic in recent exams – be prepared!

## Worked example

Find $\displaystyle\int 2\tan x \sec^5 x\,dx$     **(3 marks)**

$$\int 2\tan x \sec^5 x\,dx = \int 2\tan x \sec x\,(\sec x)^4\,dx$$

Try $y = \sec^5 x$

$$\frac{dy}{dx} = \sec x \tan x \times 5\sec^4 x$$

$$= 5\tan x \sec^5 x$$

So $\displaystyle\int 2\tan x \sec^5 x\,dx = \frac{2}{5}\sec^5 x + c$

---

## Now try this

1. Find $\displaystyle\int \frac{x+1}{x^2+2x-3}\,dx$     **(3 marks)**

2. Find $\displaystyle\int \frac{2\cos 3x}{\sin 3x}\,dx$     **(3 marks)**

     Try $y = \ln(2e^x - 1)$

3. The function f is defined by

$$f : x \mapsto \frac{e^x}{2e^x-1}, \quad x \geqslant 0$$

Find the area enclosed by the curve with equation $y = f(x)$, the line $x = 1$ and the coordinate axes. Give your answer correct to 3 decimal places.     **(4 marks)**

# Integrating partial fractions

You can use the integration rules you revised on pages 36 and 37 to integrate PARTIAL FRACTIONS. You can revise writing expressions in partial fractions on page 29.

Once it is written in partial fractions, you can integrate the expression TERM-BY-TERM.

$$\frac{4x^3 + 5}{x(2x-1)^2} = 1 + \frac{5}{x} + \frac{-8}{2x-1} + \frac{11}{(2x-1)^2}$$

Write this as $11(2x-1)^{-2}$ to integrate it. The result is $-\frac{11}{2}(2x-1)^{-1}$

This integrates to $5\ln|x|$    This integrates to $-4\ln|2x-1|$

So $\displaystyle\int \frac{4x^3+5}{x(2x-1)^2}\,dx = x + 5\ln|x| - 4\ln|2x-1| - \frac{11}{2}(2x-1)^{-1} + c$

## Worked example

(a) Express $\dfrac{5x+3}{(2x-3)(x+2)}$ in partial fractions. **(3 marks)**

$$\frac{5x+3}{(2x-3)(x+2)} = \frac{A}{2x-3} + \frac{B}{x+2}$$

$$5x + 3 = A(x+2) + B(2x-3)$$

Let $x = \frac{3}{2}$:    $5\left(\frac{3}{2}\right) + 3 = A\left(\frac{3}{2} + 2\right)$

$$\underline{A = 3}$$

Let $x = -2$:    $5(-2) + 3 = B(2(-2) - 3)$

$$\underline{B = 1}$$

So $\dfrac{5x+3}{(2x-3)(x+2)} = \dfrac{3}{2x-3} + \dfrac{1}{x+2}$

The question says 'hence' so you need to use your partial fractions from part (a) to work out the integration. If you are doing **definite integration**, make sure you write out the integral before doing any substitutions. You can get method marks even if you make a mistake in your working.

(b) Hence find the exact value of $\displaystyle\int_2^6 \frac{5x+3}{(2x-3)(x+2)}\,dx$, giving your answer as a single logarithm. **(5 marks)**

$$\int_2^6 \left(\frac{3}{2x-3} + \frac{1}{x+2}\right)dx = \left[\frac{3}{2}\ln|2x-3| + \ln|x+2|\right]_2^6$$

$$= \left(\frac{3}{2}\ln 9 + \ln 8\right) - \left(\frac{3}{2}\ln 1 + \ln 4\right)$$

$$= \ln 27 + \ln 8 - \ln 1 - \ln 4$$

$$= \ln 54$$

Use the **laws of logs** to simplify your answer:

$\frac{3}{2}\ln 9 = \ln 9^{\frac{3}{2}} = \ln\left(\sqrt{9}\right)^3 = \ln 27$

Remember that $\ln a + \ln b = \ln ab$ and $\ln a - \ln b = \ln \frac{a}{b}$ so:

$\ln 27 + \ln 8 - \ln 1 - \ln 4 = \ln\left(\dfrac{27 \times 8}{1 \times 4}\right)$

$$= \ln 54$$

## Now try this

$$f(x) = \frac{18x^2 + 10}{9x^2 - 1} = A + \frac{B}{3x+1} + \frac{C}{3x-1}$$

$9x^2 - 1$ is a difference of two squares. Factorise it using $(a^2 - b^2) = (a+b)(a-b)$.

(a) Find the values of the constants $A$, $B$ and $C$. **(4 marks)**

(b) Hence find $\displaystyle\int f(x)\,dx$ **(3 marks)**

(c) Find $\displaystyle\int_1^2 f(x)\,dx$, giving your answer in the form $2 + \ln k$ where $k$ is a constant to be found. **(3 marks)**

Use the laws of logs to simplify your expression for part (c). Remember that $k$ doesn't have to be an integer.

# Identities in integration

You can use the trigonometric identities from C3 to simplify integrations. If you're not sure which identity to use, have a look at the standard integrals in the formulae booklet. See if you can write the integral in terms of one of these functions.

You can write the integral in terms of $\sec^2$ using the identity

$$\sec^2 \theta \equiv 1 + \tan^2 \theta$$

Have a look at page 12 for identities involving $\sec^2$ and $\csc^2$.

## Worked example

Find $\int \tan^2\left(\frac{x}{2}\right) dx$     **(2 marks)**

$$\int \tan^2\left(\frac{x}{2}\right) dx = \int \left(\sec^2\left(\frac{x}{2}\right) - 1\right) dx$$

$$= 2\tan\left(\frac{x}{2}\right) - x + c$$

Use this result from the formulae booklet with $k = \frac{1}{2}$:

| $f(x)$ | $\int f(x)\,dx$ |
|---|---|
| $\sec^2 kx$ | $\frac{1}{k}\tan kx$ |

## Worked example

Show that $\int_{\frac{\pi}{4}}^{0} \sin^2 x\, dx = \frac{2 + 3\pi}{8}$     **(5 marks)**

$$\int_{\frac{\pi}{4}}^{\pi} \sin^2 x\, dx = \int_{\frac{\pi}{4}}^{\pi} \left(\frac{1}{2} - \frac{1}{2}\cos 2x\right) dx$$

$$= \left[\frac{1}{2}x - \frac{1}{4}\sin 2x\right]_{\frac{\pi}{4}}^{\pi}$$

$$= \left(\frac{\pi}{2} - \frac{1}{4}\sin 2\pi\right) - \left(\frac{\pi}{8} - \frac{1}{4}\sin\frac{\pi}{2}\right)$$

$$= \frac{3\pi}{8} + \frac{1}{4}$$

$$= \frac{2 + 3\pi}{8}$$

### $\sin^2 x$ and $\cos^2 x$

You can integrate $\sin^2 x$ and $\cos^2 x$ using the double angle formulae for cos:

**1**   $\cos 2A \equiv 2\cos^2 A - 1$

**2**   $\cos 2A \equiv 1 - 2\sin^2 A$

Have a look at page 15 for a reminder about these identities.

Use identity 2 from the box above to write $\sin^2 x$ in terms of $\cos 2x$. The question says 'show that', so make sure you clearly show the integrated function before substituting your limits and evaluating the integral.

## Now try this

1. Find the exact value of $\int_{0}^{\frac{\pi}{12}} \sin 3x \cos 3x\, dx$     **(5 marks)**

Use the identity
$$\sin 2A \equiv 2\sin A \cos A$$
with $A = 3x$

2. (a) By writing $\sin 7x$ as $\sin(4x + 3x)$ and by writing $\sin x$ as $\sin(4x - 3x)$, show that
$$\sin 7x + \sin x \equiv 2\sin 4x \cos 3x$$     **(4 marks)**

(b) Hence, or otherwise, find
$$\int \sin 4x \cos 3x\, dx$$     **(2 marks)**

Use the addition formulae on page 14 and add together the two expressions. Part (b) says 'Hence' so you can save a lot of time by using your answer to part (a) to simplify the integral.

# Integration by substitution

You can use a substitution to turn a complicated integral into a simpler one. If you have to use a substitution in your C4 exam it will usually be given with the question.

Follow these steps to find $\int \dfrac{e^{3x}}{(e^{3x}+1)^2}\,dx$ using the substitution $u = e^{3x}$:

**1**   Find $\dfrac{du}{dx}$ and write an expression for $dx$ in the form $f(x)\,du$

$u = e^{3x}$   so   $\dfrac{du}{dx} = 3e^{3x}$

This is an expression for $dx$ in the form $f(x)\,du$.

$dx$ is EQUIVALENT to $\left(\dfrac{1}{3e^{3x}}\right)du$

**2**   Swap $dx$ for $f(x)\,du$ in the integral and simplify if possible.

$$\int \frac{e^{3x}}{(e^{3x}+1)^2}\,dx = \int \frac{e^{3x}}{(e^{3x}+1)^2}\left(\frac{1}{3e^{3x}}\right)du = \int \frac{1}{3(e^{3x}+1)^2}\,du$$

**3**   Substitute every $x$ to get an integral only involving $u$ and $du$.

$$= \int \frac{1}{3(u+1)^2}\,du = \int \tfrac{1}{3}(u+1)^{-2}\,du$$

The new integral should be easier to find.

**4**   Integrate with respect to $u$.

$$= -\tfrac{1}{3}(u+1)^{-1} + c$$

**5**   Use your substitution in reverse to get an answer in terms of $x$.

$$= -\tfrac{1}{3}(e^{3x}+1)^{-1} + c$$

---

## Worked example

Use the substitution $x = \sin\theta$ to find the exact value of $\displaystyle\int_0^{\frac{1}{2}} \frac{1}{(1-x^2)^{\frac{3}{2}}}\,dx$    **(7 marks)**

$x = \sin\theta$ so $\dfrac{dx}{d\theta} = \cos\theta$, so $dx = \cos\theta\,d\theta$

When $x = \frac{1}{2}$, $\theta = \arcsin\frac{1}{2} = \dfrac{\pi}{6}$

When $x = 0$, $\theta = \arcsin 0 = 0$

$$\int_0^{\frac{1}{2}} \frac{1}{(1-x^2)^{\frac{3}{2}}}\,dx = \int_0^{\frac{\pi}{6}} \frac{1}{(1-\sin^2\theta)^{\frac{3}{2}}}\cos\theta\,d\theta$$

$$= \int_0^{\frac{\pi}{6}} \frac{1}{(\cos^2\theta)^{\frac{3}{2}}}\cos\theta\,d\theta$$

$$= \int_0^{\frac{\pi}{6}} \sec^2\theta\,d\theta$$

$$= \Big[\tan\theta\Big]_0^{\frac{\pi}{6}} = \frac{\sqrt{3}}{3}$$

## Transforming limits

When you use substitution for a DEFINITE INTEGRAL you need to use the substitution to TRANSFORM your limits.

Transform limits from values of $x$ to values of $\theta$.

$$\int_{x=0}^{x=\frac{1}{2}} \frac{1}{(1-x^2)^{\frac{3}{2}}}\,dx \qquad \int_{\theta=0}^{\theta=\frac{\pi}{6}} \sec^2\theta\,d\theta$$

You can now use your values of $\theta$ with the integrated expression to evaluate the integral.

You might need to use trigonometric identities in an integration by substitution. You can use:

$\sin^2\theta + \cos^2\theta \equiv 1$ to write the whole integral in terms of $\cos\theta$.

$$\frac{1}{(\cos^2\theta)^{\frac{3}{2}}}\cos\theta \equiv \frac{1}{\cos^3\theta}\cos\theta \equiv \frac{1}{\cos^2\theta}$$

---

## Now try this

**1.** Use the substitution $u = 3 + \sin x$ to show that

$$\int \frac{\sin 2x}{(3+\sin x)^2}\,dx = 2\ln(3+\sin x) + \frac{6}{3+\sin x} + c$$

where $c$ is a constant.    **(5 marks)**

**2.** Use the substitution $u^2 = 2x + 1$ to find the exact value of

$$\int_0^4 \frac{4x}{\sqrt{2x+1}}\,dx \qquad \textbf{(7 marks)}$$

You will need to use implicit differentiation to find the relationship between $du$ and $dx$. Have a look at page 33 for a reminder.

# Integration by parts

You can integrate some functions written as a PRODUCT of two functions using integration by parts. The rule is given in the C4 section of the formulae booklet:

$$\int u\frac{dv}{dx}dx = uv - \int v\frac{du}{dx}dx$$

$u\frac{dv}{dx}$ is the function you want to integrate. You have to work out which part of the function to set as $u$ and which part to set as $\frac{dv}{dx}$.

### Choosing $u$ and $\frac{dv}{dx}$

Follow these rules for choosing which parts of the function to set as $u$ and $\frac{dv}{dx}$ in your C4 exam:

**1** ALWAYS set $\ln x$ as $u$

**2** If there's no $\ln x$, set $x$ or $x^2$ as $u$

**3** Set $e^x$, $\sin x$ or $\cos x$ as $\frac{dv}{dx}$

---

Always write down which part of the function you are setting as $u$ and which part you are setting as $\frac{dv}{dx}$. Then calculate $\frac{du}{dx}$ and $v$ and write them down before substituting. You don't need to include a constant of integration when you work out $v$ but you should include one at the end of your final integral.

You can integrate $\ln x$ by writing it as
$$\int (\ln x)(1)\,dx$$
You **always** set $u = \ln x$, so $\frac{du}{dx} = \frac{1}{x}$. This means that the second part of the formula becomes:
$$\int (x)\left(\frac{1}{x}\right)dx = \int 1\,dx$$

## Worked example

(a) Use integration by parts to find $\int x \sin 3x\,dx$    **(3 marks)**

$u = x$ $\qquad\qquad \frac{dv}{dx} = \sin 3x$

$\frac{du}{dx} = 1$ $\qquad v = -\frac{1}{3}\cos 3x$

$\int x\sin 3x\,dx = (x)\left(-\frac{1}{3}\cos 3x\right) - \int\left(-\frac{1}{3}\cos 3x\right)(1)\,dx$

$\qquad\qquad = -\frac{1}{3}x\cos 3x + \frac{1}{3}\int\cos 3x\,dx$

$\qquad\qquad = -\frac{1}{3}x\cos 3x + \frac{1}{9}\sin 3x + c$

(b) Using your answer to part (a), find $\int x^2\cos 3x\,dx$    **(3 marks)**

$u = x^2$ $\qquad\qquad \frac{dv}{dx} = \cos 3x$

$\frac{du}{dx} = 2x$ $\qquad v = \frac{1}{3}\sin 3x$

$\int x^2\cos 3x\,dx = (x^2)\left(\frac{1}{3}\sin 3x\right) - \int\left(\frac{1}{3}\sin 3x\right)(2x)\,dx$

$\qquad\qquad = \frac{1}{3}x^2\sin 3x - \frac{2}{3}\int x\sin 3x\,dx$

$\qquad\qquad = \frac{1}{3}x^2\sin 3x - \frac{2}{3}\left(-\frac{1}{3}x\cos 3x + \frac{1}{9}\sin 3x\right) + c$

$\qquad\qquad = \frac{1}{3}x^2\sin 3x + \frac{2}{9}x\cos 3x - \frac{2}{27}\sin 3x + c$

---

## Worked example

Find the exact value of $\int_1^3 \ln x\,dx$    **(4 marks)**

$u = \ln x$ $\qquad \frac{dv}{dx} = 1$

$\frac{du}{dx} = \frac{1}{x}$ $\qquad v = x$

$\int_1^3 \ln x\,dx = \left[x\ln x\right]_1^3 - \int_1^3 1\,dx$

$\qquad\qquad = \left[x\ln x - x\right]_1^3$

$\qquad\qquad = (3\ln 3 - 3) - (1\ln 1 - 1)$

$\qquad\qquad = 3\ln 3 - 2$

## Now try this

1. Use integration by parts to find
$$\int \frac{1}{x^2}\ln x\,dx$$    **(4 marks)**

2. (a) Find $\int x e^x\,dx$    **(3 marks)**

   (b) Hence show that
$$\int_0^1 x^2 e^x\,dx = e - 2$$    **(4 marks)**

You always set $e^x$ as $\frac{dv}{dx}$

# Areas and parametric curves

You can use integration to find the area under a PARAMETRIC curve. It's almost always easier to integrate WITH RESPECT TO THE PARAMETER, $t$. For a reminder about parametric equations have a look at page 30.

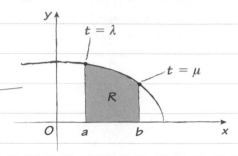

$$R = \int_{x=a}^{x=b} y\,dx = \int_{t=\lambda}^{t=\mu} y\frac{dx}{dt}dt$$

## Worked example

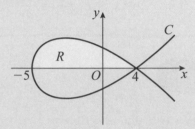

The curve $C$ with parametric equations

$$x = 3t^2 - 5 \qquad y = t(3 - t^2)$$

cuts the $x$-axis at $(-5, 0)$ and $(4, 0)$.

The region $R$ is enclosed by the curve and the $x$-axis. Use integration to find the area of $R$ in the form $k\sqrt{3}$ where $k$ is a constant.    **(5 marks)**

When $x = -5$, $t = 0$

When $x = 4$, $t = \sqrt{3}$

$\dfrac{dx}{dt} = 6t$  so $dx = 6t\,dt$

$$R = \int_{x=-5}^{x=4} y\,dx = \int_{t=0}^{t=\sqrt{3}} t(3 - t^2)(6t)\,dt$$

$$= \int_{t=0}^{t=\sqrt{3}} (18t^2 - 6t^4)\,dt$$

$$= \left[6t^3 - \frac{6}{5}t^5\right]_0^{\sqrt{3}}$$

$$= 6(\sqrt{3})^3 - \frac{6}{5}(\sqrt{3})^5$$

$$= \frac{36}{5}\sqrt{3}$$

## EXAM ALERT!

When you evaluate a definite integral with respect to the parameter, $t$, you must **transform the limits** into values of $t$. The first steps of your working should be writing down the values of $t$ at the limits, and finding $\dfrac{dx}{dt}$ so you can write down the relationship between $dx$ and $dt$.

> **Students have struggled with this topic in recent exams – be prepared!**

When you've been practising lots of complicated integrations, don't get caught out by an easy one. The expression in $t$ is just a polynomial, so multiply out and integrate.

You can write your limits as '$x = ...$' or '$t = ...$' so you know which is which. Make sure your limits **match** your operator ($dx$ or $dt$) before evaluating the integral.

For part (b) you will need to use this result from the formulae booklet:

| $f(x)$ | $\int f(x)\,dx$ |
|---|---|
| $\sec x$ | $\ln|\sec x + \tan x|$ |

## Now try this

The diagram shows the curve with parametric equations

$$x = \tan t \qquad y = 2\cos t - 1 \qquad -\frac{\pi}{2} < t < \frac{\pi}{2}$$

(a) Find the coordinates of the points $A$ and $B$ where the curve cuts the coordinate axes.    **(3 marks)**

The region $R$ is bounded by the curve, and the $x$- and $y$-axes.

(b) Use integration to find the area of $R$. Give your answer correct to 3 decimal places.    **(4 marks)**

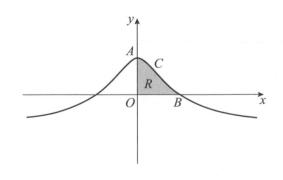

# Volumes of revolution 1

When the region $R$ is rotated through 360° around the $x$-axis, the volume of the solid formed is given by

$$V = \pi \int_a^b y^2 \, dx$$

You have to square the WHOLE of the function, $y$.

Learn this rule because it's not in the formulae booklet.

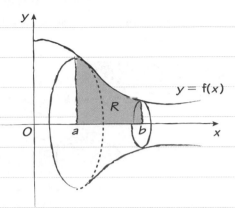

## Worked example

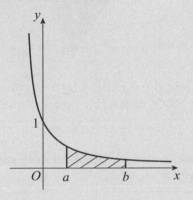

The curve shown has equation $y = \dfrac{1}{(2x + 1)}$

The finite region bounded by the curve, the $x$-axis and the lines $x = a$ and $x = b$ is shown shaded in the diagram. This region is rotated through 360° about the $x$-axis to generate a solid of revolution.

Find the volume of the solid generated.

Express your answer as a single simplified fraction, in terms of $a$ and $b$. **(5 marks)**

$$V = \pi \int_a^b y^2 \, dx = \pi \int_a^b \frac{1}{(2x+1)^2} \, dx$$

$$= \pi \int_a^b (2x+1)^{-2} \, dx$$

$$= \pi \left[ -\tfrac{1}{2}(2x+1)^{-1} \right]_a^b$$

$$= \frac{\pi}{2} \left( \frac{-1}{(2b+1)} - \frac{-1}{(2a+1)} \right)$$

$$= \frac{\pi}{2} \left( \frac{-2a - 1 + 2b + 1}{(2a+1)(2b+1)} \right)$$

$$= \frac{\pi(b-a)}{(2a+1)(2b+1)}$$

## EXAM ALERT!

Always read the question carefully. Make sure you know whether you are trying to find the **area** under the graph, or the **volume** of the solid of revolution. The only volumes you will need to find in your C4 exam are 360° rotations around the $x$-axis, so you will be able to use the formula given above.

> Students have struggled with this topic in recent exams – be prepared!

Carry out the same steps as if $a$ and $b$ were numbers, then simplify your fraction. The main things to remember with volumes of revolution are:
- Square the **whole** of $y$
- Multiply the whole integral by $\pi$.

## Now try this

The diagram shows the region $R$ which is bounded by the $x$-axis, the line $x = \dfrac{\pi}{2}$ and the curve with equation $y = 5 \sin x$. The region is rotated through $2\pi$ radians about the $x$-axis. Find the exact volume of the solid generated. **(6 marks)**

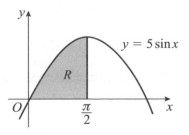

You can use $\cos 2A \equiv 1 - 2\sin^2 A$.

# Volumes of revolution 2

You can use the techniques you revised on pages 42 and 43 to find volumes of revolution generated by PARAMETRIC curves.

## Worked example

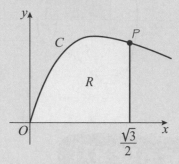

The diagram shows a sketch of the curve $C$ with parametric equations

$$x = \tfrac{1}{2}\tan t \qquad y = t\cos t \qquad 0 \le t < \frac{\pi}{2}$$

The finite region $R$ is bounded by the curve, the coordinate axes and the line $x = \dfrac{\sqrt{3}}{2}$. This region is rotated $2\pi$ radians about the $x$-axis. Find the exact volume of the solid of revolution formed, giving your answer in the form $k\pi^4$ where $k$ is a rational constant to be determined. **(7 marks)**

At point $P$, $\dfrac{\sqrt{3}}{2} = \tfrac{1}{2}\tan t$, so $t = \arctan\sqrt{3} = \dfrac{\pi}{3}$

At the origin, $0 = \tfrac{1}{2}\tan t$, so $t = \arctan 0 = 0$

$\dfrac{dx}{dt} = \tfrac{1}{2}\sec^2 t$  so  $dx = \tfrac{1}{2}\sec^2 t\, dt$

$V = \pi\displaystyle\int_{x=0}^{x=\frac{\sqrt{3}}{2}} y^2\, dx = \pi\int_{t=0}^{t=\frac{\pi}{3}}(t\cos t)^2\left(\tfrac{1}{2}\sec^2 t\right)dt$

$\qquad = \dfrac{\pi}{2}\displaystyle\int_{t=0}^{t=\frac{\pi}{3}} t^2(\cos^2 t)(\sec^2 t)\, dt$

$\qquad = \dfrac{\pi}{2}\displaystyle\int_{t=0}^{t=\frac{\pi}{3}} t^2\, dt$

$\qquad = \dfrac{\pi}{2}\left[\dfrac{1}{3}t^3\right]_0^{\frac{\pi}{3}} = \dfrac{\pi}{6}\left(\dfrac{\pi}{3}\right)^3 = \dfrac{1}{162}\pi^4$

You can substitute the parametric equation for $y$ directly into $\pi\displaystyle\int_a^b y^2\, dx$. You need to remember to:
- **transform** the limits from values of $x$ to values of $t$
- square the **whole** of the parametric equation for $y$
- find the relationship between $dx$ and $dt$ and swap the operator $dx$ for $f'(t)\, dt$.

## Cones

You can find more complicated volumes of revolution by adding or subtracting volumes of cones.

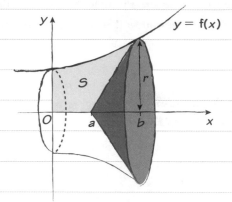

The volume of revolution formed when $S$ is rotated $360°$ about the $x$-axis is given by

$$\pi\int_0^b y^2\, dx - \frac{1}{3}\pi r^2(b-a)$$

The volume of a cone with base radius $r$ and height $h$ is $V = \frac{1}{3}\pi r^2 h$.

## Now try this

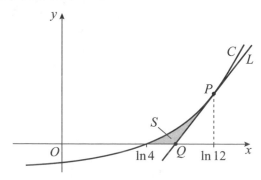

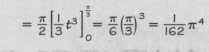

Start by finding $\dfrac{dy}{dx}$.

The curve $C$ has parametric equations

$$x = \ln 2t \qquad y = t - 2 \qquad t > 0$$

The line $L$ is a tangent to the curve at the point $P$ with coordinates $(\ln 12, 4)$.

(a) Determine the $x$-coordinate of the point $Q$ where $L$ cuts the $x$-axis. Give your answer in the form $\ln 12 - k$ where $k$ is a rational constant. **(5 marks)**

(b) The finite region $S$ is bounded by the curve, the $x$-axis and the line $L$. The region is rotated through $360°$ about the $x$-axis. Find the volume of the solid of revolution formed. **(7 marks)**

# The trapezium rule

You might have to use the trapezium rule in your C4 exam.

## Worked example

(a) Given that $y = \sec x$, complete the table with the values of $y$ corresponding to $x = \dfrac{\pi}{16}, \dfrac{\pi}{8}$ and $\dfrac{\pi}{4}$.    **(2 marks)**

| $x$ | 0 | $\dfrac{\pi}{16}$ | $\dfrac{\pi}{8}$ | $\dfrac{3\pi}{16}$ | $\dfrac{\pi}{4}$ |
|---|---|---|---|---|---|
| $y$ | 1 | 1.01959 | 1.08239 | 1.20269 | 1.41421 |

(b) Use the trapezium rule, with all the values for $y$ in the completed table, to obtain an estimate for $\displaystyle\int_0^{\frac{\pi}{4}} \sec x \, dx$. Show all the steps of your working and give your answer to 4 decimal places.    **(3 marks)**

$n = 4, \; a = 0, \; b = \dfrac{\pi}{4}, \; h = \dfrac{\pi}{16}$

$\displaystyle\int_0^{\frac{\pi}{4}} \sec x \, dx \approx \dfrac{1}{2} \times \dfrac{\pi}{16}\Big[(1 + 1.41421) +$

$\qquad\qquad 2(1.01959 + 1.08239 + 1.20269)\Big]$

$= 0.8859 \; (4 \text{ d.p.})$

The exact value of $\displaystyle\int_0^{\frac{\pi}{4}} \sec x \, dx$ is $\ln(1 + \sqrt{2})$.

(c) Calculate the % error in using the estimate you obtained in part (b).    **(2 marks)**

$\dfrac{0.8859 - \ln(1 + \sqrt{2})}{\ln(1 + \sqrt{2})} \times 100\% = 0.51\% \; (2 \text{ d.p.})$

## EXAM ALERT!

Be careful if you are asked to work out the error in an estimate.

- 'Error' or 'Absolute error'

$\qquad = \text{Estimate} - \text{Exact value}$

- 'Percentage error'

$\qquad = \dfrac{\text{Estimate} - \text{Exact value}}{\text{Exact value}} \times 100\%$

Make sure that you divide by the **exact value** when calculating a percentage error.

> Students have struggled with this topic in recent exams – be prepared!

Remember to change the width of each strip ($h$) for part (b)(ii).

---

The value for $\dfrac{3\pi}{16}$ is given to 5 d.p. so make sure you give your answers to the same degree of accuracy.

## Increasing accuracy

The trapezium rule becomes more accurate as you increase the number of strips used.

The tops of the trapezia are CLOSER to the curve so the estimate is MORE ACCURATE.

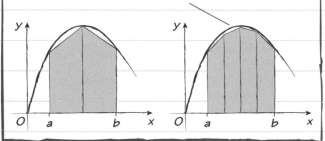

Look in the C2 section of the formulae booklet for the trapezium rule.

## Now try this

The diagram shows a sketch of the curve with equation $y = 3x^2 \ln x, \quad x > 0$.

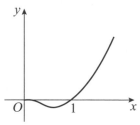

(a) Complete this table of values with the values of $y$ corresponding to $x = 1.25$, $x = 1.5$ and $x = 1.75$    **(2 marks)**

| $x$ | 1 | 1.25 | 1.5 | 1.75 | 2 |
|---|---|---|---|---|---|
| $y$ | 0 | | | | 8.31777 |

(b) Given that $I = \displaystyle\int_1^2 3x^2 \ln x \, dx$, use the trapezium rule

   (i) with values of $y$ at $x = 1$, $x = 1.5$ and $x = 2$ to find an estimate of $I$

   (ii) with all the values from the table to find another estimate of $I$.    **(5 marks)**

(c) Explain why increasing the number of values improves the accuracy of your estimate.    **(1 mark)**

45

# Solving differential equations

You revised how to FORM differential equations on page 35. You can use your integration skills to SOLVE differential equations:

**1** Separate the variables onto different sides. → **2** Integrate both sides. Only include ONE constant of integration. → **3** Substitute the boundary conditions to find the value of c. → **4** Rearrange the solution into the form needed.

## Worked example

Water is being heated in a kettle. At time $t$ seconds, the temperature of the water is $\theta\,°C$.
The rate of increase of the temperature of the water at any time $t$ is modelled by the differential equation

$$\frac{d\theta}{dt} = \lambda(120 - \theta), \qquad \theta \le 100$$

where $\lambda$ is a positive constant.
Given that $\theta = 20$ when $t = 0$,

(a) solve this differential equation to show that
$$\theta = 120 - 100e^{-\lambda t} \qquad \textbf{(8 marks)}$$

$$\int \frac{1}{120 - \theta}\, d\theta = \int \lambda\, dt$$

$$-\ln(120 - \theta) = \lambda t + c$$

When $t = 0$, $\theta = 20$:

$$-\ln(120 - 20) = \lambda(0) + c$$
$$c = -\ln 100$$

So $-\ln(120 - \theta) = \lambda t - \ln 100$
$$\ln(120 - \theta) = -\lambda t + \ln 100$$
$$120 - \theta = e^{-\lambda t + \ln 100}$$
$$= 100e^{-\lambda t}$$
$$\theta = 120 - 100e^{-\lambda t}$$

When the temperature of the water reaches 100 °C, the kettle switches off.

(b) Given that $\lambda = 0.01$, find the time, to the nearest second, when the kettle switches off. **(3 marks)**

$$100 = 120 - 100e^{-0.01t}$$
$$e^{-0.01t} = 0.2$$
$$-0.01t = \ln 0.2 = -1.6094...$$
$$t = 161 \text{ (nearest second)}$$

### Separating variables

You can't integrate expressions containing a mixture of different variables. Before you integrate both sides of a differential equation you need to separate the variables.

If $\dfrac{dy}{dx} = f(x)\,g(y)$   then   $\displaystyle\int \frac{1}{g(y)}\, dy = \int f(x)\, dx$

For example:
$$\frac{dy}{dx} = 2x^2(1 - 5y)^3 \rightarrow \int \frac{1}{(1 - 5y)^3}\, dy = \int 2x^2\, dx$$

 Remember that $\lambda$ is a constant. You can keep it on either side of the equation when you separate variables. It's easier to leave it on the 'dt' side then integrate to get $\lambda t + c$.

 Once you have integrated you have found a **general solution** to the differential equation. You could rearrange it at this point:

$$-\ln(120 - \theta) = \lambda t + c$$
$$120 - \theta = e^{-\lambda t - c} = Ae^{-\lambda t}$$
$$\theta = 120 - Ae^{-\lambda t}$$

$A$ is a constant equal to $e^{-c}$. You could substitute the boundary conditions at this stage to find the **particular solution** asked for in the question.

Separate the variables to get $\displaystyle\int \frac{1}{x}\, dx = \int \cos 2t\, dt$

## Now try this

**1.** (a) Find a general solution to the differential equation $(\sec 2t)\dfrac{dx}{dt} = x$ **(5 marks)**

(b) Find a particular solution to this equation given that $x = 2$ when $t = \dfrac{\pi}{4}$ **(2 marks)**

**2.** (a) Find $\displaystyle\int (2y + 1)^{-3}\, dy$ **(2 marks)**

(b) Given that $y = 0.5$ at $x = -8$, solve the differential equation $\dfrac{dy}{dx} = \dfrac{(2y + 1)^3}{x^2}$, giving your answer in the form $y = f(x)$. **(6 marks)**

# Vectors

In your C4 exam, three-dimensional vectors can be described using column vectors, or using **i**, **j**, **k** notation:

$$\overrightarrow{XY} = \begin{pmatrix} 3 \\ -1 \\ 4 \end{pmatrix} = 3\mathbf{i} - \mathbf{j} + 4\mathbf{k}$$

**i**, **j** and **k** are PERPENDICULAR UNIT VECTORS.

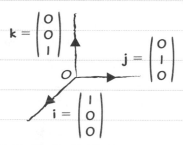

$$\mathbf{k} = \begin{pmatrix} 0 \\ 0 \\ 1 \end{pmatrix} \qquad \mathbf{j} = \begin{pmatrix} 0 \\ 1 \\ 0 \end{pmatrix} \qquad \mathbf{i} = \begin{pmatrix} 1 \\ 0 \\ 0 \end{pmatrix}$$

## Position or direction?

It is useful to distinguish between POSITION VECTORS and DIRECTION VECTORS.

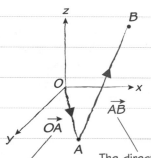

A position vector starts at the ORIGIN. $\overrightarrow{OA}$ tells you the position of point A.

The direction vector $\overrightarrow{AB}$ tells you the direction and distance from A to B.

### Magnitude

You can find the magnitude of a vector using Pythagoras' theorem.

$$\left| \overrightarrow{AB} \right| = \left| \begin{pmatrix} 2 \\ -4 \\ 5 \end{pmatrix} \right| = \left| 2\mathbf{i} - 4\mathbf{j} + 5\mathbf{k} \right|$$

$$= \sqrt{2^2 + 4^2 + 5^2} = 3\sqrt{5}$$

Ignore minus signs when calculating the magnitude of a vector.

✓ UNIT VECTORS have magnitude 1.

✓ The DISTANCE between two points A and B is the magnitude of the vector $\overrightarrow{AB}$.

## Worked example

The points $P$ and $Q$ have position vectors $3\mathbf{i} + 4\mathbf{j} + 2\mathbf{k}$ and $-\mathbf{i} + 5\mathbf{j} - 6\mathbf{k}$ respectively.

(a) Find the vector $\overrightarrow{PQ}$. **(2 marks)**

$$\overrightarrow{PQ} = \overrightarrow{OQ} - \overrightarrow{OP}$$
$$= (-1 - 3)\mathbf{i} + (5 - 4)\mathbf{j} + (-6 - 2)\mathbf{k}$$
$$= -4\mathbf{i} + \mathbf{j} - 8\mathbf{k}$$

(b) Find the distance $PQ$. **(1 mark)**

$$\left| \overrightarrow{PQ} \right| = \sqrt{4^2 + 1^2 + 8^2}$$
$$= \sqrt{81} = 9$$

(c) Find a unit vector in the direction of $PQ$. **(1 mark)**

$$\tfrac{1}{9}\overrightarrow{PQ} = \tfrac{1}{9}(-4\mathbf{i} + \mathbf{j} - 8\mathbf{k}) = -\tfrac{4}{9}\mathbf{i} + \tfrac{1}{9}\mathbf{j} - \tfrac{8}{9}\mathbf{k}$$

$$\overrightarrow{PQ} = \begin{pmatrix} \text{Position} \\ \text{vector of } Q \end{pmatrix} - \begin{pmatrix} \text{Position} \\ \text{vector of } P \end{pmatrix}$$

You could also use column vectors to subtract:

$$\begin{pmatrix} -1 \\ 5 \\ -6 \end{pmatrix} - \begin{pmatrix} 3 \\ 4 \\ 2 \end{pmatrix} = \begin{pmatrix} -1 - 3 \\ 5 - 4 \\ -6 - 2 \end{pmatrix} = \begin{pmatrix} -4 \\ 1 \\ -8 \end{pmatrix}$$

### EXAM ALERT!

You can't write in bold in your exam. You can underline vectors to make them clearer. If you're writing the vector between two points, you should draw an arrow over the top. $\overrightarrow{PQ}$ is the **direction vector** from $P$ to $Q$, whereas $PQ$ is the **line segment** between $P$ and $Q$.

Students have struggled with this topic in recent exams – be prepared!

## Now try this

1. The points $A$ and $B$ have position vectors $4\mathbf{i} - 2\mathbf{j} - 7\mathbf{k}$ and $5\mathbf{i} + 2\mathbf{j} - \mathbf{k}$ respectively.

   (a) Find the vector $\overrightarrow{AB}$. **(2 marks)**

   (b) Write down the vector $\overrightarrow{BA}$. **(1 mark)**

2. Find a unit vector in the direction of

   $$\begin{pmatrix} 2 \\ -10 \\ -11 \end{pmatrix}$$ **(2 marks)**

# Vector equations of lines

If you know the POSITION VECTOR of one point on a line and the DIRECTION VECTOR of the line you can write a vector equation for the line.

The point $P$ with position vector $\mathbf{a}$ lies on line $L$.

$L$ is parallel to the direction vector $\mathbf{b}$. A vector equation for the line is:

$$\mathbf{r} = \mathbf{a} + \lambda\mathbf{b}$$

$\mathbf{r}$ is the position vector of any point on the line.

$\lambda$ is a scalar parameter. It is different for different points along the line.

The magnitude of the direction vector doesn't affect the line produced.

## Worked example

Relative to a fixed origin, $O$, the point $A$ has position vector $8\mathbf{i} - 2\mathbf{j} + \mathbf{k}$ and the point $B$ has position vector $3\mathbf{i} + 4\mathbf{j} - 5\mathbf{k}$. The line $L$ passes through points $A$ and $B$.
Find a vector equation for the line $L$.     **(4 marks)**

$$\overrightarrow{AB} = \begin{pmatrix} -5 \\ 6 \\ -6 \end{pmatrix}$$

$$\mathbf{r} = \begin{pmatrix} 8 \\ -2 \\ 1 \end{pmatrix} + \lambda \begin{pmatrix} -5 \\ 6 \\ -6 \end{pmatrix}$$

Make sure you are confident switching between column notation and $\mathbf{i}$, $\mathbf{j}$, $\mathbf{k}$ notation. It's usually easier to work in column notation when dealing with vector equations of lines.

Make sure you write your answer as an equation. It needs to start with '$\mathbf{r} = \ldots$'.

You can start with **either** position vector, but make sure the parameter is multiplied by the **direction vector**.

## EXAM ALERT!

You can think of this as three separate equations:

$x$-coordinates: $r_x = 12 + 4\lambda$

$y$-coordinates: $r_y = 6 - \lambda$

$z$-coordinates: $r_z = -16 + 5\lambda$

In a vector equation, the value of the parameter is **different** for different points on the line. Use the coordinates you know to find the parameter at each point, then use it to calculate the unknown values. At point $A$, $\lambda = -1$, and at point $B$, $\lambda = 2$.

Students have struggled with this topic in recent exams – be prepared!

## Worked example

The line $L$ has vector equation
$$\mathbf{r} = \begin{pmatrix} 12 \\ 6 \\ -16 \end{pmatrix} + \lambda \begin{pmatrix} 4 \\ -1 \\ 5 \end{pmatrix}$$
The point $A$ has coordinates $(8, 7, a)$ and the point $B$ has coordinates $(b, 4, -6)$, where $a$ and $b$ are constants. $A$ and $B$ lie on the line $L$.
Find the values of $a$ and $b$.     **(3 marks)**

Point A: $12 + 4\lambda = 8$ so $\lambda = -1$

So  $a = -16 + 5 \times (-1) = -21$

Point B: $6 - \lambda = 4$ so $\lambda = 2$

So  $b = 12 + 2 \times 4 = 20$

## Now try this

1. With respect to a fixed origin, $O$, the point $P$ has position vector $2\mathbf{i} - 6\mathbf{j} + 9\mathbf{k}$ and the point $Q$ has position vector $5\mathbf{i} + \mathbf{j}$. The line $L$ passes through points $P$ and $Q$.

   (a) Find $\overrightarrow{PQ}$.     **(2 marks)**

   (b) Find a vector equation for $L$.     **(2 marks)**

2. The point $A$ with coordinates $(4, a, 0)$ lies on the line $L$ with vector equation
   $$\mathbf{r} = (10\mathbf{i} + 8\mathbf{j} - 12\mathbf{k}) + \lambda(\mathbf{i} - \mathbf{j} + b\mathbf{k})$$

   (a) Find the values of $a$ and $b$.     **(3 marks)**

   The point $X$ lies on $L$ where $\lambda = -1$.

   (b) Find the coordinates of $X$.     **(1 mark)**

# Intersecting lines

Follow these steps to determine whether or not two lines intersect, and to find the point of intersection:

$\lambda$ and $\mu$ are the PARAMETERS in the vector equations of the lines.

```
Write the equations        Write three              Try to solve           No solutions    Lines DO NOT intersect.
in column notation    →    linear equations    →    the first two equations          ↗
and set them equal         involving λ and μ.        simultaneously.    Solutions                Do these              No
to each other.                                                          exist         →    solutions satisfy the   →
                                                                                            third equation?
                                                                                                    Yes
                                                                                                     ↓
```

Lines DO intersect. Substitute your values for $\lambda$ and $\mu$ into the equation of one of the lines to find the point of intersection. You can use the other equation to check.

## Worked example

With respect to a fixed origin $O$, the lines $L_1$ and $L_2$ are given by the equations

$L_1: \mathbf{r} = (-6\mathbf{i} + 11\mathbf{k}) + \lambda(\mathbf{i} - \mathbf{j} + \mathbf{k})$

$L_2: \mathbf{r} = (2\mathbf{i} - 2\mathbf{j} + 9\mathbf{k}) + \mu(2\mathbf{i} + \mathbf{j} - 3\mathbf{k})$

Show that $L_1$ and $L_2$ meet and find the position vector of their point of intersection.    **(6 marks)**

$$\begin{pmatrix} -6 \\ 0 \\ 11 \end{pmatrix} + \lambda \begin{pmatrix} 1 \\ -1 \\ 1 \end{pmatrix} = \begin{pmatrix} 2 \\ -2 \\ 9 \end{pmatrix} + \mu \begin{pmatrix} 2 \\ 1 \\ -3 \end{pmatrix}$$

$-6 + \lambda = 2 + 2\mu$    ①

$-\lambda = -2 + \mu$    ②

$11 + \lambda = 9 - 3\mu$    ③

① + ②: $-6 = 3\mu$, so $\mu = -2$

Substitute into ①: $-6 + \lambda = 2 + 2(-2)$

$\lambda = 4$

Substitute $\mu = -2$ and $\lambda = 4$ into ③:

LHS = $11 + 4 = 15$

RHS = $9 - 3(-2) = 15 =$ LHS ✓

Using $L_1$, point of intersection is:

$$\begin{pmatrix} -6 \\ 0 \\ 11 \end{pmatrix} + 4 \begin{pmatrix} 1 \\ -1 \\ 1 \end{pmatrix} = \begin{pmatrix} -2 \\ -4 \\ 15 \end{pmatrix}$$

## EXAM ALERT!

Be careful with intersection questions. If you are **told** that the lines intersect you only need to solve **two** simultaneous equations to find the point of intersection. If you have to **show that** the lines intersect, you need to solve two simultaneous equations **and check** that the solution satisfies the third equation. You can check your answer by plugging your values of $\mu$ and $\lambda$ into **both** equations.

**Check it!**

Using $L_2$: $\begin{pmatrix} 2 \\ -2 \\ 9 \end{pmatrix} - 2 \begin{pmatrix} 2 \\ 1 \\ -3 \end{pmatrix} = \begin{pmatrix} -2 \\ -4 \\ 15 \end{pmatrix}$ ✓

Students have struggled with this topic in recent exams – be prepared!

## Collinear points

To determine whether three points, $A$, $B$ and $C$, are COLLINEAR (lie on the same straight line):

✓ Find the vector equation of the line through ANY TWO of the points.

✓ Check whether the THIRD point lies on the line.

## Now try this

The line $L_1$ has equation $\mathbf{r} = \begin{pmatrix} 3 \\ 1 \\ -2 \end{pmatrix} + \lambda \begin{pmatrix} 2 \\ 2 \\ 3 \end{pmatrix}$ and the line $L_2$ has equation $\mathbf{r} = \begin{pmatrix} 5 \\ 4 \\ 0 \end{pmatrix} + \mu \begin{pmatrix} 2 \\ 1 \\ -1 \end{pmatrix}$

(a) Show that $L_1$ and $L_2$ do not meet.    **(4 marks)**

(b) $A$ is the point on $L_1$ with $\lambda = 1$ and $B$ is the point on $L_2$ with $\mu = 2$.
   $C$ is the point with position vector $(-7\mathbf{i} - 6\mathbf{j} + 10\mathbf{k})$. Show that $A$, $B$ and $C$ are collinear.    **(3 marks)**

# Scalar product

You can use the scalar product of two vectors to find unknown angles.

$$\mathbf{a} = x_1\mathbf{i} + y_1\mathbf{j} + z_1\mathbf{k} \qquad \mathbf{b} = x_2\mathbf{i} + y_2\mathbf{j} + z_2\mathbf{k}$$

$$= \begin{pmatrix} x_1 \\ y_1 \\ z_1 \end{pmatrix} \qquad\qquad = \begin{pmatrix} x_2 \\ y_2 \\ z_2 \end{pmatrix}$$

The scalar product, $\mathbf{a} \cdot \mathbf{b}$, of $\mathbf{a}$ and $\mathbf{b}$ is given by

$$\mathbf{a} \cdot \mathbf{b} = x_1 x_2 + y_1 y_2 + z_1 z_2$$

It is called the SCALAR product because the answer is a number, not a vector.

### Golden rule

If the angle between vectors $\mathbf{a}$ and $\mathbf{b}$ is $\theta$:

$$\cos\theta = \frac{\mathbf{a} \cdot \mathbf{b}}{|\mathbf{a}||\mathbf{b}|}$$

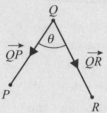

$\cos\theta > 0$, acute angle    $\cos\theta < 0$, obtuse angle

## Worked example

The points $P$, $Q$ and $R$ are such that

$$\overrightarrow{OP} = \begin{pmatrix} -5 \\ 12 \\ 6 \end{pmatrix}, \overrightarrow{OQ} = \begin{pmatrix} -2 \\ 0 \\ 2 \end{pmatrix} \text{ and } \overrightarrow{OR} = \begin{pmatrix} 8 \\ 2 \\ -9 \end{pmatrix}$$

Given that angle $PQR$ is $\theta$, show that $\cos\theta = -\frac{10}{39}$

**(4 marks)**

$$\overrightarrow{QP} = \overrightarrow{OP} - \overrightarrow{OQ} = \begin{pmatrix} -5 - (-2) \\ 12 - 0 \\ 6 - 2 \end{pmatrix} = \begin{pmatrix} -3 \\ 12 \\ 4 \end{pmatrix}$$

$$\overrightarrow{QR} = \overrightarrow{OR} - \overrightarrow{OQ} = \begin{pmatrix} 8 - (-2) \\ 2 - 0 \\ -9 - 2 \end{pmatrix} = \begin{pmatrix} 10 \\ 2 \\ -11 \end{pmatrix}$$

$$\cos\theta = \frac{\overrightarrow{QP} \cdot \overrightarrow{QR}}{|\overrightarrow{QP}||\overrightarrow{QR}|} = \frac{(-3)(10) + (12)(2) + (4)(-11)}{\sqrt{3^2 + 12^2 + 4^2}\sqrt{10^2 + 2^2 + 11^2}}$$

$$= \frac{-50}{(13)(15)} = -\frac{10}{39}$$

Use brackets when you are calculating a scalar product so you don't make a mistake with the negative signs.

When two non-perpendicular lines intersect they create an obtuse angle **and** an acute angle. Using the scalar product might generate **either** angle depending on the direction vectors in the equations.

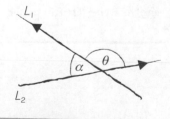

## EXAM ALERT!

Make sure both vectors are pointing away from the angle. A sketch can help.

You need to find the direction vectors $\overrightarrow{QP}$ and $\overrightarrow{QR}$, then use the golden rule above to find $\cos\theta$. The question says 'Show that ...' so make sure you write down the rule before substituting in your values.

> Students have struggled with this topic in recent exams – be prepared!

## Worked example

With respect to a fixed origin, the lines $L_1$ and $L_2$ are defined by the vector equations

$$L_1 : \mathbf{r} = (5\mathbf{i} + 2\mathbf{j} - 12\mathbf{k}) + \lambda(3\mathbf{i} - \mathbf{j} + \mathbf{k})$$

$$L_2 : \mathbf{r} = (14\mathbf{i} + 15\mathbf{j} - 16\mathbf{k}) + \mu(\mathbf{i} + 5\mathbf{j} - 2\mathbf{k})$$

Given that $L_1$ and $L_2$ intersect, find the acute angle between $L_1$ and $L_2$. **(4 marks)**

$$\cos\theta = \frac{(3)(1) + (-1)(5) + (1)(-2)}{\sqrt{3^2 + 1^2 + 1^2}\sqrt{1^2 + 5^2 + 2^2}}$$

$$= -0.22019\ldots$$

$$\theta = 102.7203\ldots°$$

Acute angle, $\alpha = 180° - 102.7203\ldots°$

$$= 77.3° \text{ (1 d.p.)}$$

## Now try this

Relative to a fixed origin, $O$, points $A$, $B$ and $C$ have position vectors $(5\mathbf{i} - 2\mathbf{j} + 10\mathbf{k})$, $(-\mathbf{i} + 6\mathbf{k})$ and $(8\mathbf{i} + 2\mathbf{j} + \mathbf{k})$ respectively. Find the size of angle $CAB$.

**(4 marks)**

# Perpendicular vectors

The scalar product (which you can revise on the previous page) makes it easy to identify PERPENDICULAR vectors. You should always use DIRECTION VECTORS, not position vectors, when finding perpendicular lines.

## Golden rule

If two vectors are perpendicular, their scalar product is 0, and vice versa.

$$\mathbf{a} \cdot \mathbf{b} = 0$$

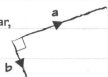

---

## Worked example

Relative to a fixed origin, $O$, the line $L$ has vector equation

$$\mathbf{r} = \begin{pmatrix} 10 \\ -1 \\ 8 \end{pmatrix} + \lambda \begin{pmatrix} -2 \\ 1 \\ 3 \end{pmatrix}$$

The point $A$ lies on $L$ where $\lambda = 3$. The point $B$ has position vector $(3p\mathbf{i} + p\mathbf{j})$ relative to $O$, where $p$ is a constant.

Given that $\overrightarrow{AB}$ is perpendicular to $L$, find the value of $p$. **(5 marks)**

$$\overrightarrow{OA} = \begin{pmatrix} 10 \\ -1 \\ 8 \end{pmatrix} + 3\begin{pmatrix} -2 \\ 1 \\ 3 \end{pmatrix} = \begin{pmatrix} 4 \\ 2 \\ 17 \end{pmatrix}$$

$$\overrightarrow{AB} = \begin{pmatrix} 3p \\ p \\ 0 \end{pmatrix} - \begin{pmatrix} 4 \\ 2 \\ 17 \end{pmatrix} = \begin{pmatrix} 3p-4 \\ p-2 \\ -17 \end{pmatrix}$$

$\overrightarrow{AB}$ is perpendicular to $L$ so

$$\begin{pmatrix} -2 \\ 1 \\ 3 \end{pmatrix} \begin{pmatrix} 3p-4 \\ p-2 \\ -17 \end{pmatrix} = 0$$

$$(-2)(3p-4) + (1)(p-2) + (3)(-17) = 0$$

$$-5p = 45 \text{ so } p = -9$$

There are lots of steps here so **plan** your answer.

1. Find the **position vector** of $A$ by substituting $\lambda = 3$ into the equation of the line.
2. Work out $\overrightarrow{OB} - \overrightarrow{OA}$ to find an expression for $\overrightarrow{AB}$ in **terms of $p$**.
3. Find an expression for the scalar product of $\overrightarrow{AB}$ and the **direction vector** of the line.
4. Set this scalar product equal to 0 and solve the equation to find $p$.

**Check it!**

If $p = -9$ then $\overrightarrow{OB} = \begin{pmatrix} -27 \\ -9 \\ 0 \end{pmatrix}$ and $\overrightarrow{AB} = \begin{pmatrix} -31 \\ -11 \\ -17 \end{pmatrix}$

So $\overrightarrow{AB} \cdot \begin{pmatrix} -2 \\ 1 \\ 3 \end{pmatrix} = (-31)(-2) + (-11)(1) + (-17)(3)$

$= 62 - 11 - 51 = 0$ ✔

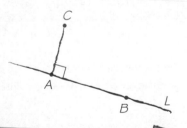

A sketch can help you visualise a vectors question.

---

## Shortest distance

You can often use basic trigonometry to find the shortest distance from a point to a line.

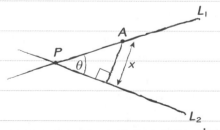

The length of the line segment $PA$ is $|\overrightarrow{PA}|$, so the shortest distance from $A$ to $L_2$ is

$$x = |\overrightarrow{PA}| \sin\theta$$

## Now try this

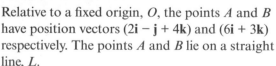

Relative to a fixed origin, $O$, the points $A$ and $B$ have position vectors $(2\mathbf{i} - \mathbf{j} + 4\mathbf{k})$ and $(6\mathbf{i} + 3\mathbf{k})$ respectively. The points $A$ and $B$ lie on a straight line, $L$.

(a) Find a vector equation for the line $L$. **(4 marks)**

The point $C$ has position vector $(p\mathbf{i} - 3\mathbf{j} + 2p\mathbf{k})$ with respect to $O$, where $p$ is a constant.

Given that $AC$ is perpendicular to $L$, find

(b) the value of $p$ **(4 marks)**

(c) the perpendicular distance of the point $C$ from $L$. **(2 marks)**

# Solving area problems

You can use these formulae to find areas of TRIANGLES and PARALLELOGRAMS in vector questions:

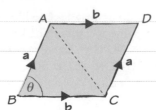

**①** Area of triangle ABC = $\frac{1}{2}|\mathbf{a}||\mathbf{b}|\sin\theta$

The area of the parallelogram is TWICE the area of the triangle.

**②** Area of parallelogram ABCD = $|\mathbf{a}||\mathbf{b}|\sin\theta$

## Worked example

Three points $A$, $B$ and $C$ are such that

$$\overrightarrow{AB} = \begin{pmatrix} 3 \\ 2 \\ 6 \end{pmatrix} \text{ and } \overrightarrow{AC} = \begin{pmatrix} 4 \\ -2 \\ 0 \end{pmatrix}$$

Find the area of the triangle $ABC$. **(4 marks)**

$|\overrightarrow{AB}| = \sqrt{3^2 + 2^2 + 6^2} = 7$

$|\overrightarrow{AC}| = \sqrt{4^2 + 2^2 + 0^2} = 2\sqrt{5}$

$\cos\angle BAC = \dfrac{\overrightarrow{AB} \cdot \overrightarrow{AC}}{|\overrightarrow{AB}||\overrightarrow{AC}|}$

$= \dfrac{(3)(4) + (2)(-2) + (6)(0)}{(7)(2\sqrt{5})}$

$= 0.2555...$

$\angle BAC = 75.1937...$

Area $= \frac{1}{2}|\overrightarrow{AB}||\overrightarrow{AC}|\sin\angle BAC$

$= \frac{1}{2} \times 7 \times 2\sqrt{5} \times \sin(75.1937...)$

$= 15.13$ (2 d.p.)

You can draw a sketch to help you find the right lengths and angles.

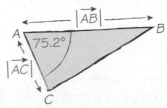

You will need to use $|\overrightarrow{AB}|$ and $|\overrightarrow{AC}|$ to calculate the angle $BAC$ **and** to find the area of the triangle. Work them out first and leave them as exact answers to avoid making rounding errors later.

## Now try this

1. Points $A$, $B$ and $C$ have coordinates $(5, -1, 0)$, $(2, 4, 10)$ and $(6, -1, 4)$ respectively.

   (a) Find the vectors $\overrightarrow{CA}$ and $\overrightarrow{CB}$. **(2 marks)**

   (b) Find the area of the triangle $ABC$. **(4 marks)**

   (c) Point $D$ is such that the points $A$, $B$, $C$ and $D$ form the vertices of a parallelogram. Find the coordinates of three possible positions of $D$. **(5 marks)**

   (d) Write down the area of the parallelogram. **(1 mark)**

2. The line $L_1$ has equation $\mathbf{r} = \begin{pmatrix} 1 \\ -2 \\ 4 \end{pmatrix} + \lambda\begin{pmatrix} 6 \\ 0 \\ 1 \end{pmatrix}$, where $\lambda$ is a scalar parameter.

   The line $L_2$ has equation $\mathbf{r} = \begin{pmatrix} 5 \\ 12 \\ 0 \end{pmatrix} + \mu\begin{pmatrix} 2 \\ -2 \\ 1 \end{pmatrix}$, where $\mu$ is a scalar parameter.

   (a) Given that $L_1$ and $L_2$ meet at the point $P$, find the coordinates of $P$. **(3 marks)**

   $Q$ is the point on $L_1$ where $\lambda = 0$ and $R$ is the point on $L_2$ where $\mu = 5$.

   (b) Find the size of the angle $QPR$, giving your answer in degrees to 2 decimal places. **(4 marks)**

   (c) Hence, or otherwise, find the area of the triangle $PQR$. **(5 marks)**

Use direction vectors to find the possible positions of D. There are three possibilities.

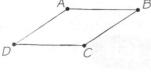
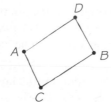
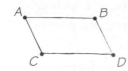

# You are the examiner!

CHECKING YOUR WORK is one of the key skills you will need for your C4 exam. All five of these students have made ONE key mistake in their working. Can you spot them all?

**1**

$$\frac{9x^2}{(x-1)^2(2x+1)} = \frac{A}{(x-1)} + \frac{B}{(x-1)^2} + \frac{C}{(2x+1)}$$

Find the values of the constants $A$, $B$ and $C$.  **(4 marks)**

$9x^2 = A(x-1)(2x+1) + B(2x+1) + C(x-1)^2$

Let $x = 1$:        $9 = 3B$

$\underline{B = 3}$

Let $x = -\frac{1}{2}$:    $\frac{9}{4} = -\frac{3}{2}^2 C$

$\frac{9}{4} = -\frac{9}{4}C$

$\underline{C = -1}$

Equate $x^2$ terms:   $9 = 2A + C$

$9 = 2A - 1$

$\underline{A = 5}$

**2** Find $\int x \cos 3x \, dx$      **(4 marks)**

$u = x$        $\frac{dv}{dx} = \cos 3x$

$\frac{du}{dx} = 1$      $v = -3 \sin 3x$

$\int x \cos 3x \, dx = uv - \int v \frac{du}{dx} dx$

$= -3x \sin 3x + \int 3 \sin 3x \, dx$

$= -3x \sin 3x - 9 \cos 3x + c$

**3** A curve $C$ has equation $x^3 + 2y^2 - 4xy = 1$

Find $\frac{dy}{dx}$ in terms of $x$ and $y$.    **(4 marks)**

$3x^2 + 4y\frac{dy}{dx} - 4y - 4x\frac{dy}{dx} = 1$

$\frac{dy}{dx}(4y - 4x) = 4y - 3x^2 + 1$

$\frac{dy}{dx} = \frac{4y - 3x^2 + 1}{4y - 4x}$

**4** Relative to a fixed origin, $O$, the points $A$ and $B$ have position vectors $(6\mathbf{i} - 3\mathbf{j} + \mathbf{k})$ and $(\mathbf{i} - 2\mathbf{k})$ respectively. The line $L$ passes through $A$ and $B$. Find a vector equation for the line $L$.    **(3 marks)**

$\overrightarrow{AB} = \begin{pmatrix} 1 \\ 0 \\ -2 \end{pmatrix} - \begin{pmatrix} 6 \\ -3 \\ 1 \end{pmatrix} = \begin{pmatrix} -5 \\ 3 \\ -3 \end{pmatrix}$

$\begin{pmatrix} 6 \\ -3 \\ 1 \end{pmatrix} + \lambda \begin{pmatrix} -5 \\ 3 \\ -3 \end{pmatrix}$

**5** Use the substitution $u = x - 1$ to find the exact value of $\int_2^4 \frac{x}{\sqrt{x-1}} dx$    **(5 marks)**

$u = x - 1$ so $\frac{du}{dx} = 1$, so $du \equiv dx$

$\int_2^4 \frac{x}{\sqrt{x-1}} dx = \int_2^4 \frac{u+1}{\sqrt{u}} du = \int_2^4 (u^{\frac{1}{2}} + u^{-\frac{1}{2}}) du$

$= \left[ \frac{2}{3}u^{\frac{3}{2}} + 2u^{\frac{1}{2}} \right]_2^4$

$= \left( \frac{16}{3} + 4 \right) - \left( \frac{4\sqrt{2}}{3} + 2\sqrt{2} \right)$

$= \frac{2}{3}(14 - 5\sqrt{2})$

## Checking your work

If you have any time left at the end of your exam, you should check back through your working.

☑ Check you have answered EVERY PART and given all the information asked for, in the CORRECT FORM.

☑ Make sure everything is EASY TO READ.

☑ Cross out any incorrect working with a SINGLE NEAT LINE and UNDERLINE the correct answer.

☑ DOUBLE CHECK any complicated algebraic working, especially when BRACKETS and NEGATIVE NUMBERS are involved.

**Now try this**

Find the mistake in each student's answer on this page, and write out the correct working for each question. Turn over for the answers.

# You are still the examiner!

BEFORE looking at this page, turn back to page 53 and try to spot the key mistake in each student's working. Use this page to CHECK your answers — the corrections are shown in red, and these answers are now 100% CORRECT.

 **1**

$$\frac{9x^2}{(x-1)^2(2x+1)} = \frac{A}{(x-1)} + \frac{B}{(x-1)^2} + \frac{C}{(2x+1)}$$

Find the values of the constants $A$, $B$ and $C$.

**(4 marks)**

$9x^2 = A(x-1)(2x+1) + B(2x+1) + C(x-1)^2$

Let $x = 1$: $\qquad 9 = 3B$

$\qquad\qquad\qquad \underline{B = 3}$

Let $x = -\frac{1}{2}$: $\qquad \frac{9}{4} = (-\frac{3}{2})^2 C$

$\qquad\qquad\qquad \frac{9}{4} = +\frac{9}{4}C$

$\qquad\qquad\qquad \underline{C = +1}$

Equate $x^2$ terms: $\quad 9 = 2A + C$

$\qquad\qquad\qquad 9 = 2A + 1$

$\qquad\qquad\qquad \underline{A = \cancel{5}\ 4}$

**Top tip**

Use **brackets** when dealing with negative numbers.

Revise partial fractions on page 29.

 **2** Find $\int x \cos 3x \,dx$    **(4 marks)**

$u = x \qquad\qquad \frac{dv}{dx} = \cos 3x$

$\frac{du}{dx} = 1 \qquad\qquad v = \cancel{-3\sin 3x} = \frac{1}{3}\sin 3x$

$\int x \cos 3x \,dx = uv - \int v \frac{du}{dx}\,dx$

$\qquad = \cancel{-3x\sin 3x + \int 3\sin 3x\,dx}$

$\qquad = \cancel{-3x\sin 3x - 9\cos 3x + c}$

$\qquad = \frac{1}{3}x\sin 3x - \int \frac{1}{3}\sin 3x\,dx$

$\qquad = \frac{1}{3}x\sin 3x + \frac{1}{9}\cos 3x + c$

**Top tip**

When you **integrate** a function of the form $f(kx)$ you need to **divide** by $k$.

Revise integration by parts on page 41.

 **3** A curve $C$ has equation $x^3 + 2y^2 - 4xy = 1$

Find $\frac{dy}{dx}$ in terms of $x$ and $y$.    **(4 marks)**

$3x^2 + 4y\frac{dy}{dx} - 4y - 4x\frac{dy}{dx} \cancel{= 1} = 0$

$\frac{dy}{dx}(4y - 4x) = 4y - 3x^2 \cancel{+1}$

$\frac{dy}{dx} = \frac{4y - 3x^2 \cancel{+1}}{4y - 4x}$

**Top tip**

You have to differentiate **every term** including the constant on the right-hand side.

For more on implicit differentiation see page 33.

**4** Relative to a fixed origin, $O$, the points $A$ and $B$ have position vectors $(6\mathbf{i} - 3\mathbf{j} + \mathbf{k})$ and $(\mathbf{i} - 2\mathbf{k})$ respectively. The line $L$ passes through $A$ and $B$. Find a vector equation for the line $L$.    **(3 marks)**

$$\overrightarrow{AB} = \begin{pmatrix} 1 \\ 0 \\ -2 \end{pmatrix} - \begin{pmatrix} 6 \\ -3 \\ 1 \end{pmatrix} = \begin{pmatrix} -5 \\ 3 \\ -3 \end{pmatrix}$$

$$\mathbf{r} = \begin{pmatrix} 6 \\ -3 \\ 1 \end{pmatrix} + \lambda \begin{pmatrix} -5 \\ 3 \\ -3 \end{pmatrix}$$

**Top tip**

Always read vectors questions carefully. You need to find an **equation** of the line. This means you need to write '$\mathbf{r} = \dots$' at the start.

Revise this on page 48.

 **5** Use the substitution $u = x - 1$ to find the exact value of $\int_2^4 \frac{x}{\sqrt{x-1}}\,dx$    **(5 marks)**

$u = x - 1$ so $\frac{du}{dx} = 1$, so $du \equiv dx$

$\int_2^4 \frac{x}{\sqrt{x-1}}\,dx = \int_{\cancel{2}\,1}^{\cancel{4}\,3} \frac{u+1}{\sqrt{u}}\,du = \int_{\cancel{2}\,1}^{\cancel{4}\,3} (u^{\frac{1}{2}} + u^{-\frac{1}{2}})\,du$

$\qquad = \left[ \frac{2}{3}u^{\frac{3}{2}} + 2u^{\frac{1}{2}} \right]_{\cancel{2}\,1}^{\cancel{4}\,3}$

$\qquad = (2\sqrt{3} + 2\sqrt{3}) - (\frac{2}{3} + 2) \cancel{= (\frac{16}{3} + 4) - (\frac{4\sqrt{2}}{3} + 2\sqrt{2})}$

$\qquad = 4\sqrt{3} - \frac{8}{3} \qquad\qquad \cancel{= \frac{2}{3}(14 - 5\sqrt{2})}$

**Top tip**

For **definite integration** using substitution, you need to change the limits to values of $u$ before calculating.

Integration by substitution is covered on page 40.

# Worked solutions

## CORE MATHEMATICS 3

### 1 Algebraic fractions

1. $\dfrac{3x^2 - 8x - 3}{x^2 - 9} = \dfrac{(3x + 1)\cancel{(x - 3)}}{(x + 3)\cancel{(x - 3)}}$

$= \dfrac{3x + 1}{x + 3}$

2. $\dfrac{x + 5}{2x^2 + 7x - 4} - \dfrac{1}{2x - 1} = \dfrac{x + 5}{(2x - 1)(x + 4)} - \dfrac{1}{2x - 1}$

$= \dfrac{x + 5}{(2x - 1)(x + 4)} - \dfrac{x + 4}{(2x - 1)(x + 4)}$

$= \dfrac{x + 5 - (x + 4)}{(2x - 1)(x + 4)} = \dfrac{1}{(2x - 1)(x + 4)}$

### 2 Algebraic division

$2x^4 + 4x^2 - x + 2 \equiv (x^2 - 1)(ax^2 + bx + c) + dx + e$

$\equiv ax^4 + bx^3 + cx^2 - ax^2 - bx - c + dx + e$

$\equiv ax^4 + bx^3 + (c - a)x^2 + (d - b)x + (e - c)$

$x^4$ terms $\rightarrow \underline{a = 2}$

$x^3$ terms $\rightarrow \underline{b = 0}$

$x^2$ terms $\rightarrow c - a = 4$ so $\underline{c = 6}$

$x$ terms $\rightarrow d - b = -1$ so $\underline{d = -1}$

Constant terms $\rightarrow e - c = 2$ so $\underline{e = 8}$

### 3 Functions

(a) $f(-2) = 3(-2) + 2 = -6 + 2 = -4$

$gf(-2) = g(-4) = \dfrac{-4}{-4 + 2} = \dfrac{-4}{-2} = 2$

(b) $fg(x) = 3\left(\dfrac{x}{x + 2}\right) + 2$

$= \dfrac{3x}{x + 2} + \dfrac{2(x + 2)}{x + 2}$

$= \dfrac{5x + 4}{x + 2}$

(c) $3x + 2 = (3x + 2)^2$

$3x + 2 = 9x^2 + 12x + 4$

$0 = 9x^2 + 9x + 2$

$0 = (3x + 1)(3x + 2)$

$x = -\frac{1}{3}$ or $x = -\frac{2}{3}$

### 4 Graphs and range

1. (a) $-6 \leqslant f(x) \leqslant 4$      (b) $ff(-1) = f(2) = 3$

2. (a) $g(x) = \dfrac{3x - 7}{x^2 - 5x + 6} - \dfrac{2}{x - 3}$

$= \dfrac{3x - 7}{(x - 2)(x - 3)} - \dfrac{2(x - 2)}{(x - 2)(x - 3)}$

$= \dfrac{3x - 7 - 2(x - 2)}{(x - 2)(x - 3)}$

$= \dfrac{\cancel{x - 3}}{(x - 2)\cancel{(x - 3)}}$

$= \dfrac{1}{x - 2}$

(b) When $x = 3$, $g(x) = \dfrac{1}{3 - 2} = 1$

As $x \to \infty$, $g(x) \to 0$ but is always $> 0$.

Range is $0 < g(x) < 1$

### 5 Inverse functions

(a)      $y = \dfrac{x + 5}{x}$

$xy = x + 5$

$xy - x = 5$

$x(y - 1) = 5$

$x = \dfrac{5}{y - 1}$

$h^{-1}(x) = \dfrac{5}{x - 1}$

(b) Range of $h^{-1}$ = domain of h, so $h^{-1}(x) > 1$

(c) When $x = 1$, $h(x) = 6$

As $x \to \infty$, $h(x) \to 1$ but is always $> 1$.

Range of h is $1 < h(x) < 6$

Domain of $h^{-1}$ is $1 < x < 6$

### 6 Inverse graphs

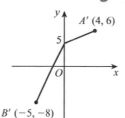

### 7 Modulus

(a) 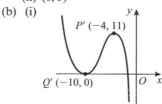      (b)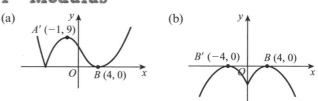

### 8 Transformations of graphs

(a) (i) $(5, -6)$

(ii) $(6, 3)$

(b) (i)

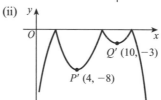

(ii)

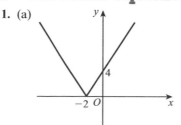

### 9 Modulus equations

1. (a)

(b) The graph of $y = x$ does not intersect the graph of $y = f(x)$ so the equation $f(x) = x$ has no solutions.

(c) $|2x + 4| = -x$

Positive argument

$2x + 4 = -x$

$x = -\frac{4}{3}$

Negative argument

$-(2x + 4) = -x$

$x = -4$

The graph of $y = -x$ would intersect the graph of $y = f(x)$ twice so both solutions exist.

**2.** $|x - 1| = 4 - 2x$

Positive argument

$x - 1 = 4 - 2x$

$x = \frac{5}{3}$

Check: $2x + 1 = 2\left(\frac{5}{3}\right) + 1 = \frac{13}{3}$

$5 - |x - 1| = 5 - \left|\frac{5}{3} - 1\right| = \frac{13}{3}$ so $x = \frac{5}{3}$ is a solution.

Negative argument

$-(x - 1) = 4 - 2x$

$x = 3$

Check: $2x + 1 = 7$

$5 - |x - 1| = 3$ so $x = 3$ is not a solution

## 10 Sec, cosec and cot

**1.**

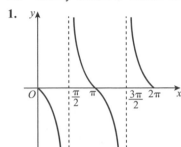

**2.**

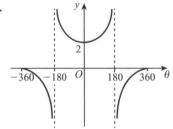

## 11 Trig equations 1

**1.** $\cot^2 x - 3\cot x + 2 = 0$

$(\cot x - 2)(\cot x - 1) = 0$

$\cot x = 2 \qquad\qquad \cot x = 1$

$\tan x = \frac{1}{2} \qquad\qquad \tan x = 1$

$x = 26.6°, 206.6° \qquad x = 45.0°, 225.0°$

**2.** $0 \leqslant x \leqslant \pi$ so $0 \leqslant 2x \leqslant 2\pi$

$\sec 2x = \frac{2}{\sqrt{3}}$

$\cos 2x = \frac{\sqrt{3}}{2}$

$2x = \frac{\pi}{6}, \frac{11\pi}{6}$

$x = \frac{\pi}{12}, \frac{11\pi}{12}$

## 12 Using trig identities

**1.** $\operatorname{cosec}^2 x - \sec^2 x \equiv (1 + \cot^2 x) - (1 + \tan^2 x)$

$\equiv \cot^2 x - \tan^2 x$

**2.** $3(\sec^2\theta - 1) + 7\sec\theta = 3$

$3\sec^2\theta + 7\sec\theta - 6 = 0$

$(3\sec\theta - 2)(\sec\theta + 3) = 0$

$\sec\theta = \frac{2}{3}$ ✗ $\qquad \sec\theta = -3$ ✓

$\cos\theta = -\frac{1}{3}$

$\theta = 109.5°, 250.5°$

## 13 Arcsin, arccos and arctan

**1.** (a) $\frac{\pi}{3}$ $\qquad$ (b) $\frac{\pi}{3}$ $\qquad$ (c) $\frac{\pi}{4}$

**2.** (a) $g\left(\frac{1}{2}\right) = \arcsin\frac{1}{2} - \frac{\pi}{4}$

$= \frac{\pi}{6} - \frac{\pi}{4}$

$= -\frac{\pi}{12}$

(b) $\arcsin x - \frac{\pi}{4} = 0$

$\arcsin x = \frac{\pi}{4}$

$x = \frac{1}{\sqrt{2}}$

(c) $y = \arcsin x - \frac{\pi}{4}$

$y + \frac{\pi}{4} = \arcsin x$

$x = \sin\left(y + \frac{\pi}{4}\right)$

$g^{-1}(x) = \sin\left(x + \frac{\pi}{4}\right), -\frac{3\pi}{4} \leqslant x \leqslant \frac{\pi}{4}$

(d)

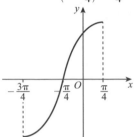

## 14 Addition formulae

**1.** $\sin 15° = \sin(45° - 30°)$

$= \sin 45° \cos 30° - \cos 45° \sin 30°$

$= \frac{1}{\sqrt{2}} \times \frac{\sqrt{3}}{2} - \frac{1}{\sqrt{2}} \times \frac{1}{2}$

$= \frac{\sqrt{3} - 1}{2\sqrt{2}}$

$= \frac{\sqrt{6} - \sqrt{2}}{4}$

$\operatorname{cosec} 15° = \frac{1}{\sin 15°} = \frac{4}{\sqrt{6} - \sqrt{2}}$

**2.** (a) $\cos(x + 45°) = \cos x \cos 45° - \sin x \sin 45°$

$= \cos x \times \frac{1}{\sqrt{2}} - \sin x \times \frac{1}{\sqrt{2}}$

$= \frac{\cos x - \sin x}{\sqrt{2}}$

(b) $\sqrt{2}\cos(x + 45°) = 0.5$

$\cos(x + 45°) = \frac{1}{2\sqrt{2}}$

$0 \leqslant x \leqslant 360°$ so $45° \leqslant x + 45° \leqslant 405°$

$x + 45° = 69.3°, 290.7°$

$x = 24.3°, 245.7°$

## 15 Double angle formulae

**1.** $\cos 2A \equiv \cos(A + A) \equiv \cos A \cos A - \sin A \sin A$

$\equiv \cos^2 A - \sin^2 A$

$\equiv \cos^2 A - (1 - \cos^2 A)$

$\equiv 2\cos^2 A - 1$

**2.** (a) $\cos 3x = \cos(2x + x)$

$= \cos 2x \cos x - \sin 2x \sin x$

$= (2\cos^2 x - 1)\cos x - (2\sin x \cos x)\sin x$

$= 2\cos^3 x - \cos x - 2\sin^2 x \cos x$

$= 2\cos^3 x - \cos x - 2(1 - \cos^2 x)\cos x$

$= 2\cos^3 x - \cos x - 2\cos x + 2\cos^3 x$

$= 4\cos^3 x - 3\cos x$

(b) $2(4\cos^3 x - 3\cos x) - 1 = 0$

$4\cos^3 x - 3\cos x = \frac{1}{2}$

$\cos 3x = \frac{1}{2}$

$0 \leqslant x \leqslant 90°$ so $0 \leqslant 3x \leqslant 270°$

$3x = 60°$

$x = 20°$

## 16 $a\cos\theta \pm b\sin\theta$

1. (a) $R = \sqrt{3^2 + 2^2} = \sqrt{13}$

$\alpha = \arctan\left(\dfrac{2}{3}\right) = 0.5880\ldots$

$3\sin 2\theta + 2\cos 2\theta = \sqrt{13}\sin(2\theta + 0.5880\ldots)$

(b) $3\sin 2\theta + 2\cos 2\theta = -1$

$\sqrt{13}\sin(2\theta + 0.5880\ldots) = -1$

$\sin(2\theta + 0.5880\ldots) = -\dfrac{1}{\sqrt{13}}$

$0 \le \theta \le \pi$ so $0.5880\ldots \le 2\theta + 0.5880\ldots \le 2\pi + 0.5880\ldots$

$2\theta + 0.5880\ldots = -0.2810\ldots, 3.4226\ldots, 6.0021\ldots$

$\theta = 1.42, 2.71$ (2 d.p.)

2. (a) $R = \sqrt{\sqrt{3}^2 + 1^2} = \sqrt{4} = 2$

$\alpha = \arctan\left(\dfrac{1}{\sqrt{3}}\right) = 30°$

(b) $y = 2\cos(x - 30°)$

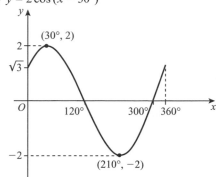

## 17 Trig modelling

(a) $R = \sqrt{4^2 + 2^2} = \sqrt{20} = 2\sqrt{5}$

$\alpha = \arctan\left(\dfrac{2}{4}\right) = 0.4636\ldots$

(b) $2\sqrt{5}$ cm

(c) $2\sqrt{5}\cos(1.2t - 0.4636) = 0$

$\cos(1.2t - 0.4636) = 0$

$0 < t < 5$ so $-0.4636 < 1.2t - 0.4636 < 5.5364$

$1.2t - 0.4636 = \dfrac{\pi}{2}, \dfrac{3\pi}{2}$

$t = 1.70, 4.31$ (2 d.p.)

## 18 Exponential functions

1. (a) Asymptote: $x = 0$

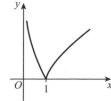

(b) Asymptote: $x = 2$

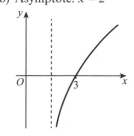

(c) Asymptote: $x = 0$

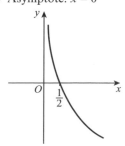

2. $6 = 3e^{2x-1}$

$2 = e^{2x-1}$

$\ln 2 = 2x - 1$

$x = \tfrac{1}{2} + \tfrac{1}{2}\ln 2$

## 19 Exponential equations

1. (a) $\ln\left(\dfrac{x+1}{x}\right) = \ln 5$

$\dfrac{x+1}{x} = 5$

$x + 1 = 5x$

$x = \dfrac{1}{4}$

(b) $(e^{2x})^2 + 3e^{2x} - 10 = 0$

$(e^{2x} + 5)(e^{2x} - 2) = 0$

$e^{2x} = -5$ or $e^{2x} = 2$

$2x = \ln 2$

$x = \tfrac{1}{2}\ln 2$

(c) $\ln(6x + 7) = \ln x^2$

$6x + 7 = x^2$

$x^2 - 6x - 7 = 0$

$(x - 7)(x + 1) = 0$

$x = 7$ or $x = -1$

2. $\ln(3^x e^{2x-5}) = \ln 7$

$\ln 3^x + \ln e^{2x-5} = \ln 7$

$x\ln 3 + 2x - 5 = \ln 7$

$x(2 + \ln 3) = 5 + \ln 7$

$x = \dfrac{5 + \ln 7}{2 + \ln 3}$

3. (a) $f(x) = \dfrac{3x^2 - 7x + 2}{x^2 - 4}$

$= \dfrac{(3x - 1)(x - 2)}{(x + 2)(x - 2)}$

$= \dfrac{3x - 1}{x + 2}$

(b) $\ln(3x^2 - 7x + 2) - \ln(x^2 - 4) = 1$

$\ln\left(\dfrac{3x^2 - 7x + 2}{x^2 - 4}\right) = 1$

$\dfrac{3x^2 - 7x + 2}{x^2 - 4} = e$

$\dfrac{3x - 1}{x + 2} = e$

$3x - 1 = ex + 2e$

$3x - ex = 1 + 2e$

$x(3 - e) = 1 + 2e$

$x = \dfrac{1 + 2e}{3 - e}$

## 20 Exponential modelling

1. (a) $4100$

(b) $8000 = 4000 + 100e^{0.8t}$

$e^{0.8t} = 40$

$0.8t = \ln 40$

$t = \dfrac{\ln 40}{0.8} = 4.6110\ldots$ hours

$= 4$ hours and $36.6$ minutes

$= 4$ hours and $37$ minutes (4.37 pm)

2. (a) $250$ grams

(b) $125 = 250e^{-90k}$

$e^{-90k} = 0.5$

$-90k = \ln 0.5$

$k = 0.00770$ (3 s.f.)

## 21 The chain rule

1. $y = (x^2 - 3x + 1)^{-\frac{1}{2}}$

$\dfrac{dy}{dx} = -\tfrac{1}{2}(x^2 - 3x + 1)^{-\frac{3}{2}} \times (2x - 3)$

$= \dfrac{3 - 2x}{2\sqrt{(x^2 - 3x + 1)^3}}$

**2.** $f(x) = \left(x^{\frac{1}{3}} + 6\right)^6$

$f'(x) = 6\left(x^{\frac{1}{3}} + 6\right)^5 \times \frac{1}{3}x^{-\frac{2}{3}}$

$\qquad = \frac{6\left(\sqrt[3]{x} + 6\right)^5}{3\left(\sqrt[3]{x}\right)^2}$

$f'(8) = \frac{6\left(\sqrt[3]{8} + 6\right)^5}{3\left(\sqrt[3]{8}\right)^2} = \frac{6(8)^5}{3(2)^2} = \frac{6(2^3)^5}{3(2)^2} = \frac{2(2^{15})}{2^2} = 2^{14}$

## 22 Derivatives to learn

**1.** (a) $2\cos 2x + 4\sin 4x$

(b) $2x - 4e^{4x-3}$

(c) $\dfrac{6x}{3x^2 + 1}$

**2.** $\dfrac{dx}{dy} = 2\cos 2y$

$\qquad = 2\sqrt{1 - \sin^2 2y}$

$\qquad = 2\sqrt{1 - x^2}$

$\dfrac{dy}{dx} = \dfrac{1}{\left(\dfrac{dx}{dy}\right)} = \dfrac{1}{2\sqrt{1 - x^2}}$

**3.** $\dfrac{d}{dx}(\sin^2 x) = \dfrac{d}{dx}(\sin x)^2$

$\qquad = 2\sin x \times \cos x$

$\qquad = \sin 2x$

## 23 The product rule

**1.** (a) $\quad u = \sqrt{x} \qquad v = \sin 5x$

$\qquad \dfrac{du}{dx} = \dfrac{1}{2\sqrt{x}} \qquad \dfrac{dv}{dx} = 5\cos 5x$

$\qquad \dfrac{d}{dx}(uv) = u\dfrac{dv}{dx} + v\dfrac{du}{dx}$

$\qquad\qquad = 5\sqrt{x}\cos 5x + \dfrac{\sin 5x}{2\sqrt{x}}$

$\qquad\qquad = \dfrac{10x\cos 5x + \sin 5x}{2\sqrt{x}}$

(b) $\quad u = x \qquad v = \ln x$

$\qquad \dfrac{du}{dx} = 1 \qquad \dfrac{dv}{dx} = \dfrac{1}{x}$

$\qquad \dfrac{d}{dx}(uv) = u\dfrac{dv}{dx} + v\dfrac{du}{dx}$

$\qquad\qquad = x\left(\dfrac{1}{x}\right) + \ln x$

$\qquad\qquad = 1 + \ln x$

**2.** (a) $\quad u = e^x \qquad v = 3x^2 - 4x - 1$

$\qquad \dfrac{du}{dx} = e^x \qquad \dfrac{dv}{dx} = 6x - 4$

$\qquad \dfrac{dy}{dx} = u\dfrac{dv}{dx} + v\dfrac{du}{dx}$

$\qquad\qquad = e^x(6x - 4) + e^x(3x^2 - 4x - 1)$

$\qquad\qquad = e^x(3x^2 + 2x - 5)$

(b) When $\dfrac{dy}{dx} = 0$, $e^x(3x^2 + 2x - 5) = 0$

$\qquad e^x \neq 0$ so $3x^2 + 2x - 5 = 0$

$\qquad\qquad (3x + 5)(x - 1) = 0$

$\qquad x = -\frac{5}{3}$ or $x = 1$

## 24 The quotient rule

**1.** (a) $\dfrac{\sqrt{x}(-\sin x) - (\cos x)\left(\frac{1}{2}x^{-\frac{1}{2}}\right)}{\left(\sqrt{x}\right)^2} = \dfrac{-2x\sin x - \cos x}{2x^{\frac{3}{2}}}$

(b) $\dfrac{(\sqrt{3x+1})(2) - (2x)\left(\frac{3}{2}\right)(3x+1)^{-\frac{1}{2}}}{\left(\sqrt{3x+1}\right)^2} = \dfrac{3x+2}{(3x+1)^{\frac{3}{2}}}$

(c) $\dfrac{(\ln(x^2+1))(2x) - (x^2+1)\left(\dfrac{2x}{x^2+1}\right)}{(\ln(x^2+1))^2} = \dfrac{2x(\ln(x^2+1) - 1)}{(\ln(x^2+1))^2}$

**2.** $\dfrac{d}{dx}(\cot x) = \dfrac{d}{dx}\left(\dfrac{\cos x}{\sin x}\right)$

$\qquad = \dfrac{(\sin x)(-\sin x) - (\cos x)(\cos x)}{(\sin x)^2}$

$\qquad = \dfrac{-(\sin^2 x + \cos^2 x)}{\sin^2 x}$

$\qquad = \dfrac{-1}{\sin^2 x}$

$\qquad = -\csc^2 x$

**3.** (a) $f(x) = \dfrac{3(x^2+2)}{(x^2+2)(x-1)} - \dfrac{3(x-1)}{(x^2+2)(x-1)} - \dfrac{9}{(x^2+2)(x-1)}$

$\qquad = \dfrac{3(x^2+2) - 3(x-1) - 9}{(x^2+2)(x-1)}$

$\qquad = \dfrac{3x(x-1)}{(x^2+2)(x-1)}$

$\qquad = \dfrac{3x}{x^2+2}$

(b) $f'(x) = \dfrac{(x^2+2)(3) - (3x)(2x)}{(x^2+2)^2}$

$\qquad = \dfrac{6 - 3x^2}{(x^2+2)^2}$

## 25 Differentiation and graphs

(a) $\dfrac{dy}{dx} = (e^{2x})(\sec^2 x) + (\tan x)(2e^{2x})$

$\qquad = e^{2x}(\sec^2 x + 2\tan x)$

$\qquad = e^{2x}(\tan^2 x + 2\tan x + 1)$

$\qquad = e^{2x}(1 + \tan x)^2$

(b) $e^{2x}(1 + \tan x)^2 = 0$

$\qquad 1 + \tan x = 0$

$\qquad \tan x = -1$

$\qquad x = -\dfrac{\pi}{4}$

When $x = -\dfrac{\pi}{4}$, $y = e^{2\left(-\frac{\pi}{4}\right)}\tan\left(-\dfrac{\pi}{4}\right)$

$\qquad\qquad = -e^{-\frac{\pi}{2}}$

Coordinates of $P$ are $\left(-\dfrac{\pi}{4}, -e^{-\frac{\pi}{2}}\right)$

(c) When $x = 1$:

$y = e^2\tan 1 = 11.5077\ldots$

$\dfrac{dy}{dx} = e^2(1 + \tan 1)^2 = 48.3268\ldots$

Equation of tangent:

$y - 11.5077\ldots = 48.3268\ldots(x - 1)$

$y = 48.3x - 36.8$

## 26 Iteration

(a) $f(1) = \ln(1 + 1) - 2(1) + 2$

$\qquad = 0.693\ldots$ Positive

$f(2) = \ln(2 + 1) - 2(2) + 2$

$\qquad = -0.901\ldots$ Negative

Change of sign so $1 < \alpha < 2$

(b) $x_1 = 1.45815$

$\quad x_2 = 1.44970$

$\quad x_3 = 1.44798$

(c) $f(1.44745) = \ln(1.44745 + 1) - 2(1.44745) + 2$

$\qquad\qquad = 0.000146\ldots$ Positive

$f(1.44755) = \ln(1.44755 + 1) - 2(1.44755) + 2$

$\qquad\qquad = -0.0000124\ldots$ Negative

Change of sign, so $1.44745 < \alpha < 1.44755$

So $\alpha = 1.4475$ correct to 4 decimal places

## CORE MATHEMATICS 4

### 29 Partial fractions

1. $\dfrac{8x^2}{(3x-2)(x+2)^2} = \dfrac{A}{(3x-2)} + \dfrac{B}{(x+2)} + \dfrac{C}{(x+2)^2}$

$8x^2 = A(x+2)^2 + B(3x-2)(x+2) + C(3x-2)$

Let $x = -2$: $8(-2)^2 = C(3(-2)-2)$

$32 = -8C$

$\underline{C = -4}$

Let $x = \frac{2}{3}$: $8\left(\frac{2}{3}\right)^2 = A\left(\frac{2}{3} + 2\right)^2$

$\frac{32}{9} = \frac{64}{9}A$

$\underline{A = \frac{1}{2}}$

Equate $x^2$ terms: $8 = A + 3B$

$8 = \frac{1}{2} + 3B$

$\underline{B = \frac{5}{2}}$

$\dfrac{8x^2}{(x+2)^2(3x-2)} = \dfrac{1}{2(3x-2)} + \dfrac{5}{2(x+2)} - \dfrac{4}{(x+2)^2}$

2. $6x^2 - 1 = A(2x-3)(x+1) + B(x+1) + C(2x-3)$

Let $x = -1$: $6(-1)^2 - 1 = C(2(-1)-3)$

$5 = -5C$

$\underline{C = -1}$

Let $x = \frac{3}{2}$: $6\left(\frac{3}{2}\right)^2 - 1 = B\left(\frac{3}{2} + 1\right)$

$\frac{25}{2} = \frac{5}{2}B$

$\underline{B = 5}$

Equate $x^2$ terms: $6 = 2A$

$\underline{A = 3}$

### 30 Parametric equations

$y = \sin t \cos \dfrac{\pi}{3} + \cos t \sin \dfrac{\pi}{3}$

$= \dfrac{1}{2}\sin t + \dfrac{\sqrt{3}}{2}\cos t$

$= \dfrac{1}{2}\sin t + \dfrac{\sqrt{3}}{2}\sqrt{1 - \sin^2 t}$

$= \dfrac{1}{2}x + \dfrac{\sqrt{3}}{2}\sqrt{1 - x^2}$

### 31 Parametric differentiation

(a) $\dfrac{dx}{dt} = -2\sin\left(t - \dfrac{\pi}{3}\right)$, $\dfrac{dy}{dt} = 6\cos 2t$

$\dfrac{dy}{dx} = \dfrac{6\cos 2t}{-2\sin\left(t - \dfrac{\pi}{3}\right)}$

(b) When $t = 0$, $x = 2\cos\left(-\dfrac{\pi}{3}\right) = 1$, $y = 3\sin 0 = 0$

$\dfrac{dy}{dx} = \dfrac{6\cos 0}{-2\sin\left(-\dfrac{\pi}{3}\right)} = \dfrac{6}{-2\left(-\dfrac{\sqrt{3}}{2}\right)} = 2\sqrt{3}$

Gradient of normal $= \dfrac{-1}{2\sqrt{3}}$

Equation of normal:

$y - 0 = \dfrac{-1}{2\sqrt{3}}(x - 1)$

$x + 2\sqrt{3}y - 1 = 0$

(c) $\dfrac{dy}{dx} = 0$ means $6\cos 2t = 0$

$\cos 2t = 0$, $0 \leqslant t < 2\pi$

$t = \dfrac{\pi}{4}, \dfrac{3\pi}{4}, \dfrac{5\pi}{4}, \dfrac{7\pi}{4}$

$t = \dfrac{\pi}{4}$: $x = 2\cos\left(\dfrac{\pi}{4} - \dfrac{\pi}{3}\right) = \dfrac{\sqrt{6} + \sqrt{2}}{2}$

$y = 3\sin 2\left(\dfrac{\pi}{4}\right) = 3$

$t = \dfrac{3\pi}{4}$: $x = 2\cos\left(\dfrac{3\pi}{4} - \dfrac{\pi}{3}\right) = \dfrac{\sqrt{6} - \sqrt{2}}{2}$

$y = 3\sin 2\left(\dfrac{3\pi}{4}\right) = -3$

$t = \dfrac{5\pi}{4}$: $x = 2\cos\left(\dfrac{5\pi}{4} - \dfrac{\pi}{3}\right) = \dfrac{-\sqrt{6} - \sqrt{2}}{2}$

$y = 3\sin 2\left(\dfrac{5\pi}{4}\right) = 3$

$t = \dfrac{7\pi}{4}$: $x = 2\cos\left(\dfrac{7\pi}{4} - \dfrac{\pi}{3}\right) = \dfrac{-\sqrt{6} + \sqrt{2}}{2}$

$y = 3\sin 2\left(\dfrac{7\pi}{4}\right) = -3$

Coordinates are $\left(\dfrac{\sqrt{6} + \sqrt{2}}{2}, 3\right)$, $\left(\dfrac{\sqrt{6} - \sqrt{2}}{2}, -3\right)$, $\left(\dfrac{-\sqrt{6} - \sqrt{2}}{2}, 3\right)$,

$\left(\dfrac{-\sqrt{6} + \sqrt{2}}{2}, -3\right)$

### 32 The binomial series

(a) $(1 + (-6x))^{\frac{1}{3}} = 1 + \left(\dfrac{1}{3}\right)(-6x) + \left(\dfrac{\left(\frac{1}{3}\right)\left(-\frac{2}{3}\right)}{1 \times 2}\right)(-6x)^2 +$

$\left(\dfrac{\left(\frac{1}{3}\right)\left(-\frac{2}{3}\right)\left(-\frac{5}{3}\right)}{1 \times 2 \times 3}\right)(-6x)^3 + \ldots$

$= 1 - 2x - 4x^2 - \dfrac{40}{3}x^3 + \ldots$

(b) $x = 0.01$

$\sqrt[3]{0.94} \approx 1 - 2(0.01) - 4(0.01)^2 - \dfrac{40}{3}(0.01)^3$

$= 0.979587$

### 33 Implicit differentiation

1. $3x^2 + 2y\dfrac{dy}{dx} + 6xy + 3x^2\dfrac{dy}{dx} = 0$

$\dfrac{dy}{dx}(2y + 3x^2) = -3x^2 - 6xy$

$\dfrac{dy}{dx} = \dfrac{-3x^2 - 6xy}{2y + 3x^2}$

At $P$, $\dfrac{dy}{dx} = \dfrac{-3(3)^2 - 6(3)(-1)}{2(-1) + 3(3)^2} = -\dfrac{9}{25}$

2. (a) LHS: $2(0)^2 - (-1)^2 = -1$

RHS: $-1\,e^{3(0)} = -1$

So $(0, -1)$ lies on $C$.

(b) $4x - 2y\dfrac{dy}{dx} = 3y\,e^{3x} + e^{3x}\dfrac{dy}{dx}$

$4x - 3y\,e^{3x} = \dfrac{dy}{dx}(2y + e^{3x})$

$\dfrac{dy}{dx} = \dfrac{4x - 3y\,e^{3x}}{2y + e^{3x}}$

At $(0, -1)$, $\dfrac{dy}{dx} = \dfrac{4(0) - 3(-1)\,e^{3(0)}}{2(-1) + e^{3(0)}} = -3$

Equation of tangent:

$y - (-1) = -3(x - 0)$

$3x + y + 1 = 0$

### 34 Differentiating $a^x$

1. $\dfrac{d}{dx}[4^x \sin x] = 4^x \cos x + 4^x \ln 4 \sin x$

$= 4^x(\cos x + \ln 4 \sin x)$

**2.** $y + x\dfrac{dy}{dx} + \left(\tfrac{1}{2}\right)^y \ln\tfrac{1}{2}\dfrac{dy}{dx} = 0$

$\dfrac{dy}{dx}\left(x + \left(\tfrac{1}{2}\right)^y \ln\tfrac{1}{2}\right) = -y$

$\dfrac{dy}{dx} = \dfrac{-y}{x + \left(\tfrac{1}{2}\right)^y \ln\tfrac{1}{2}}$

At $(0, -1)$, $\dfrac{dy}{dx} = \dfrac{-1}{0 + \left(\tfrac{1}{2}\right)^{-1}\ln 2} = \dfrac{-1}{2\ln 2}$

Gradient of normal = $2\ln 2 = \ln 4$
Equation of normal:
$\quad y - (-1) = \ln 4(x - 0)$
$(\ln 4)x - y - 1 = 0$

## 35 Rates of change

(a)

$\tan\theta = \dfrac{h}{x}$ so $x = h\cot\theta$

$V = \tfrac{1}{2}h[6 + (6 + 2h\cot\theta)] \times 40$

$V = 240h + 40h^2\cot\theta$

$\dfrac{dV}{dh} = 240 + 80h\cot\theta$

(b) $\dfrac{dh}{dt} = \dfrac{dV}{dt} \div \dfrac{dV}{dh}$

$\quad 0.1 = \dfrac{40}{240 + 80(2.5)\cot\theta}$

$24 + 20\cot\theta = 40$

$\cot\theta = 0.8$

$\tan\theta = 1.25$

$\theta = 51.3°$ (1 d.p.)

## 36 Integrals to learn

**1.** (a) $-2\cos 2x + c$ 　　　(b) $6e^{\frac{x}{6}} + c$

**2.** $\int_0^{\frac{\pi}{2}} 2\sin\tfrac{1}{2}\theta\, d\theta = \left[-4\cos\tfrac{1}{2}\theta\right]_0^{\frac{\pi}{2}}$

$\quad = \left(-4\cos\tfrac{\pi}{4}\right) - (-4\cos 0)$

$\quad = -2\sqrt{2} + 4$

**3.** $\tfrac{1}{3}\ln|\sec 3x| + c$

## 37 Reverse chain rule

**1.** $\tfrac{1}{2}\int\dfrac{2x + 2}{x^2 + 2x - 3}\,dx = \tfrac{1}{2}\ln|x^2 + 2x - 3| + c$

**2.** $\tfrac{2}{3}\int\dfrac{3\cos 3x}{\sin 3x}\,dx = \tfrac{2}{3}\ln|\sin 3x| + c$

**3.** $\int_0^1 f(x)\,dx = \tfrac{1}{2}\int_0^1\dfrac{2e^x}{2e^x - 1}\,dx$

$\quad = \tfrac{1}{2}\left[\ln|2e^x - 1|\right]_0^1$

$\quad = \tfrac{1}{2}\ln(2e - 1) - \tfrac{1}{2}\ln(1)$

$\quad = 0.745$ (3 d.p.)

## 38 Integrating partial fractions

(a) $18x^2 + 10 = A(3x + 1)(3x - 1) + B(3x - 1) + C(3x + 1)$

Let $x = \tfrac{1}{3}$: $18\left(\tfrac{1}{3}\right)^2 + 10 = C\left(3\left(\tfrac{1}{3}\right) + 1\right)$

$\quad\quad\quad 12 = 2C$

$\quad\quad\quad \underline{C = 6}$

Let $x = -\tfrac{1}{3}$: $18\left(-\tfrac{1}{3}\right)^2 + 10 = B\left(3\left(-\tfrac{1}{3}\right) - 1\right)$

$\quad\quad\quad 12 = -2B$

$\quad\quad\quad \underline{B = -6}$

Equate $x^2$ terms: $18 = 9A$

$\quad\quad\quad \underline{A = 2}$

(b) $\int\left(2 - \dfrac{6}{3x + 1} + \dfrac{6}{3x - 1}\right)dx = 2x - 2\ln|3x + 1| + 2\ln|3x - 1| + c$

$\quad\quad = 2x + \ln\left(\dfrac{3x - 1}{3x + 1}\right)^2 + c$

(c) $\int_1^2 f(x)\,dx = \left[2x + \ln\left(\dfrac{3x - 1}{3x + 1}\right)^2\right]_1^2$

$\quad = \left(4 + \ln\left(\dfrac{25}{49}\right)\right) - \left(2 + \ln\left(\dfrac{1}{4}\right)\right)$

$\quad = 2 + \ln\left(\dfrac{100}{49}\right)$

## 39 Identities in integration

**1.** $\int_0^{\frac{\pi}{12}}\sin 3x\cos 3x\,dx = \int_0^{\frac{\pi}{12}}\tfrac{1}{2}\sin 6x\,dx = \left[-\tfrac{1}{12}\cos 6x\right]_0^{\frac{\pi}{12}}$

$\quad = \left(-\tfrac{1}{12}\cos\tfrac{\pi}{2}\right) - \left(-\tfrac{1}{12}\cos 0\right)$

$\quad = \tfrac{1}{12}$

**2.** (a) $\sin 7x = \sin(4x + 3x) = \sin 4x\cos 3x + \cos 4x\sin 3x$ 　①

$\quad\quad \sin x = \sin(4x - 3x) = \sin 4x\cos 3x - \cos 4x\sin 3x$ 　②

$\quad\quad ① + ② \quad \sin 7x + \sin x = 2\sin 4x\cos 3x$

(b) $\int\sin 4x\cos 3x\,dx = \int\left(\tfrac{1}{2}\sin 7x + \tfrac{1}{2}\sin x\right)dx$

$\quad\quad = -\tfrac{1}{14}\cos 7x - \tfrac{1}{2}\cos x + c$

## 40 Integration by substitution

**1.** $u = 3 + \sin x$ so $\dfrac{du}{dx} = \cos x$ 　　and　 $dx = \left(\dfrac{1}{\cos x}\right)du$

$\int\dfrac{\sin 2x}{(3 + \sin x)^2}\,dx = \int\dfrac{\sin 2x}{u^2}\dfrac{1}{\cos x}\,du$

$= \int\dfrac{2\sin x\cos x}{u^2}\dfrac{1}{\cos x}\,du$

$= \int\dfrac{2(u - 3)}{u^2}\,du$

$= \int\left(\dfrac{2}{u} - \dfrac{6}{u^2}\right)du$

$= 2\ln|u| + 6u^{-1} + c$

$= 2\ln(3 + \sin x) + \dfrac{6}{3 + \sin x} + c$

**2.** $u^2 = 2x + 1$ so $2u\dfrac{du}{dx} = 2$

So $dx = u\,du$

When $x = 4$, $u = 3$

When $x = 0$, $u = 1$

$\int_0^4\dfrac{4x}{\sqrt{2x + 1}}\,dx = \int_1^3\dfrac{2(u^2 - 1)}{u}u\,du$

$\quad = \int_1^3(2u^2 - 2)\,du$

$\quad = \left[\tfrac{2}{3}u^3 - 2u\right]_1^3$

$\quad = \left(\tfrac{2}{3}(3)^3 - 2(3)\right) - \left(\tfrac{2}{3}(1)^3 - 2(1)\right)$

$\quad = 12 - \left(-\tfrac{4}{3}\right)$

$\quad = 13\tfrac{1}{3} = \tfrac{40}{3}$

## 41 Integration by parts

**1.** 　$u = \ln x$ 　　　　　$\dfrac{dv}{dx} = \dfrac{1}{x^2}$

$\dfrac{du}{dx} = \dfrac{1}{x}$ 　　　　　$v = -\dfrac{1}{x}$

$\int\dfrac{1}{x^2}\ln x\,dx = (\ln x)\left(-\dfrac{1}{x}\right) - \int\left(-\dfrac{1}{x}\right)\left(\dfrac{1}{x}\right)dx$

$\quad = -\dfrac{\ln x}{x} + \int\left(\dfrac{1}{x^2}\right)dx$

$\quad = -\dfrac{\ln x}{x} - \dfrac{1}{x} + c$

**2.** (a)    $u = x$      $\dfrac{dv}{dx} = e^x$

$\dfrac{du}{dx} = 1$      $v = e^x$

$\int x\,e^x\,dx = x\,e^x - \int e^x\,dx$

$= x\,e^x - e^x + c$

$= e^x(x - 1) + c$

(b)    $u = x^2$      $\dfrac{dv}{dx} = e^x$

$\dfrac{du}{dx} = 2x$      $v = e^x$

$\int_0^1 x^2 e^x\,dx = \left[x^2 e^x\right]_0^1 - 2\int_0^1 x\,e^x\,dx$

$= \left[x^2 e^x - 2e^x(x-1)\right]_0^1$

$= \left[e^x(x^2 - 2x + 2)\right]_0^1$

$= e - 2$

## 42 Areas and parametric curves

(a) When $x = 0$, $\tan t = 0$

So $t = 0$, and $y = 2\cos(0) - 1 = 1$

$A$ is the point $(0, 1)$

When $y = 0$, $2\cos t - 1 = 0$

So $t = \pm\dfrac{\pi}{3}$, and $x = \tan\pm\dfrac{\pi}{3} = \pm\sqrt{3}$

From the diagram, $B$ is the point $(\sqrt{3}, 0)$

(b) $\dfrac{dx}{dt} = \sec^2 t$ so $dx = \sec^2 t\,dt$

$R = \int_{x=0}^{x=\sqrt{3}} y\,dx = \int_{t=0}^{t=\frac{\pi}{3}} (2\cos t - 1)(\sec^2 t)\,dt$

$= \int_{t=0}^{t=\frac{\pi}{3}} (2\sec t - \sec^2 t)\,dt$

$= \left[2\ln|\sec x + \tan x| - \tan t\right]_0^{\frac{\pi}{3}}$

$= (2\ln(2 + \sqrt{3}) - \sqrt{3}) - (2\ln(1) - 0))$

$= 2\ln(2 + \sqrt{3}) - \sqrt{3} = 0.902 \text{ (3 d.p.)}$

## 43 Volumes of revolution 1

$V = \pi\int_0^{\frac{\pi}{2}} y^2\,dx = \pi\int_0^{\frac{\pi}{2}} (5\sin x)^2\,dx$

$= 25\pi\int_0^{\frac{\pi}{2}} \sin^2 x\,dx$

$= 25\pi\int_0^{\frac{\pi}{2}} \tfrac{1}{2}(1 - \cos 2x)\,dx$

$= \dfrac{25\pi}{2}\left[x - \tfrac{1}{2}\sin 2x\right]_0^{\frac{\pi}{2}}$

$= \dfrac{25\pi}{2}\left(\left(\dfrac{\pi}{2} - \tfrac{1}{2}\sin\pi\right) - \left(0 - \tfrac{1}{2}\sin 0\right)\right)$

$= \dfrac{25\pi^2}{4}$

## 44 Volumes of revolution 2

(a) $\dfrac{dy}{dx} = \dfrac{dy}{dt} \div \dfrac{dx}{dt} = 1 \div \dfrac{1}{t} = t$

At $P$, $t - 2 = 4$ so $t = 6$

Equation of $L$:

$y - 4 = 6(x - \ln 12)$

When $y = 0$, $-4 = 6x - 6\ln 12$

$6x = 6\ln 12 - 4$

$x = \ln 12 - \tfrac{2}{3}$

(b) $dx = \dfrac{1}{t}dt$

At $(\ln 4, 0)$, $t - 2 = 0$ so $t = 2$

Volume *including* cone $= \pi\int_{x=\ln 4}^{x=\ln 12} y^2\,dx = \pi\int_{t=2}^{t=6} (t-2)^2\left(\dfrac{1}{t}\right)dt$

$= \pi\int_{t=2}^{t=6} \left(t - 4 + \dfrac{4}{t}\right)dt$

$= \pi\left[\tfrac{1}{2}t^2 - 4t + 4\ln t\right]_2^6$

$= \pi\left[(-6 + 4\ln 6) - (-6 + 4\ln 2)\right]$

$= 4\pi\ln 3$

Volume of cone $= \tfrac{1}{3}\pi(4)^2\left(\tfrac{2}{3}\right) = \dfrac{32\pi}{9}$

Volume required $= 4\pi\ln 3 - \dfrac{32\pi}{9} = 2.635 \text{ (3 d.p.)}$

## 45 The trapezium rule

(a)
| $x$ | 1 | 1.25 | 1.5 | 1.75 | 2 |
|---|---|---|---|---|---|
| $y$ | 0 | 1.04599 | 2.73689 | 5.14147 | 8.31777 |

(b) (i) $I \approx \tfrac{1}{2} \times 0.5[(0 + 8.31777) + 2(2.73689)] = 3.4479 \text{ (4 d.p.)}$

(ii) $I \approx \tfrac{1}{2} \times 0.25[(0 + 8.31777) + 2(1.04599 + 2.73689 + 5.14147)] = 3.2708 \text{ (4 d.p.)}$

(c) Because the tops of the trapezia are closer to the curve

## 46 Solving differential equations

**1.** (a) $\int\frac{1}{x}\,dx = \int\cos 2t\,dt$

$\ln x = \tfrac{1}{2}\sin 2t + c$

(b) $\ln 2 = \tfrac{1}{2}\sin 2\left(\dfrac{\pi}{4}\right) + c$

$= \tfrac{1}{2} + c$

$c = \ln 2 - \tfrac{1}{2}$

$\ln x = \tfrac{1}{2}\sin 2t + \ln 2 - \tfrac{1}{2}$

**2.** (a) $\int(2y + 1)^{-3}\,dy = -\tfrac{1}{4}(2y + 1)^{-2} + c$

(b) $\int(2y + 1)^{-3}\,dy = \int\dfrac{1}{x^2}\,dx$

$-\tfrac{1}{4}(2y + 1)^{-2} = -\dfrac{1}{x} + c$

When $y = 0.5$, $x = -8$:

$-\tfrac{1}{4}(2(0.5) + 1)^{-2} = -\dfrac{1}{-8} + c$

$-\dfrac{1}{16} = \dfrac{1}{8} + c$

$c = -\dfrac{3}{16}$

$-\tfrac{1}{4}(2y + 1)^{-2} = -\dfrac{1}{x} - \dfrac{3}{16}$

$\dfrac{1}{(2y + 1)^2} = \dfrac{4}{x} + \dfrac{3}{4}$

$\dfrac{1}{(2y + 1)^2} = \dfrac{16 + 3x}{4x}$

$2y + 1 = \pm\sqrt{\dfrac{4x}{16 + 3x}}$

$y = -\dfrac{1}{2} \pm \sqrt{\dfrac{x}{16 + 3x}}$

## 47 Vectors

**1.** (a) $\overrightarrow{AB} = \overrightarrow{OB} - \overrightarrow{OA}$

$= (5 - 4)\mathbf{i} + (2 - -2)\mathbf{j} + (-1 - -7)\mathbf{k}$

$= \mathbf{i} + 4\mathbf{j} + 6\mathbf{k}$

(b) $\overrightarrow{BA} = -\mathbf{i} - 4\mathbf{j} - 6\mathbf{k}$

**2.** $\left\|\begin{pmatrix}2\\-10\\-11\end{pmatrix}\right\| = \sqrt{2^2 + 10^2 + 11^2} = 15$

Unit vector $= \dfrac{1}{15}\begin{pmatrix}2\\-10\\-11\end{pmatrix} = \begin{pmatrix}\frac{2}{15}\\-\frac{2}{3}\\-\frac{11}{15}\end{pmatrix}$

## 48 Vector equations of lines

**1.** (a) $\overrightarrow{PQ} = (5 - 2)\mathbf{i} + (1 - -6)\mathbf{j} + (0 - 9)\mathbf{k}$

$= 3\mathbf{i} + 7\mathbf{j} - 9\mathbf{k}$

(b) $\mathbf{r} = (2\mathbf{i} - 6\mathbf{j} + 9\mathbf{k}) + \lambda(3\mathbf{i} + 7\mathbf{j} - 9\mathbf{k})$

or $\mathbf{r} = (5\mathbf{i} + \mathbf{j}) + \lambda(3\mathbf{i} + 7\mathbf{j} - 9\mathbf{k})$

**2.** (a) From $\mathbf{i}$ components: $10 + \lambda = 4$ so $\lambda = -6$

From $\mathbf{j}$ components: $8 - \lambda = a$

$8 - (-6) = a$

$a = 14$

From $\mathbf{k}$ components: $-12 + \lambda b = 0$

$-12 - 6b = 0$

$b = -2$

(b) $\overrightarrow{OX} = (10\mathbf{i} + 8\mathbf{j} - 12\mathbf{k}) - (\mathbf{i} - \mathbf{j} - 2\mathbf{k})$
$= 9\mathbf{i} + 9\mathbf{j} - 10\mathbf{k}$
$X$ has coordinates $(9, 9, -10)$

## 49 Intersecting lines

(a) $\quad 3 + 2\lambda = 5 + 2\mu \qquad ①$
$\quad 1 + 2\lambda = 4 + \mu \qquad ②$
$\quad -2 + 3\lambda = -\mu \qquad ③$
$\quad ① - ② : 2 = 1 + \mu$
$\qquad\qquad\quad \mu = 1$
From ②: $1 + 2\lambda = 5$
$\qquad\qquad \lambda = 2$
Check using ③:
LHS: $-2 + 3(2) = 4$
RHS: $-1$
Not equal so lines do *not* intersect

(b) $A = \begin{pmatrix} 5 \\ 3 \\ 1 \end{pmatrix}$, $B = \begin{pmatrix} 9 \\ 6 \\ -2 \end{pmatrix}$

$\overrightarrow{AB} = \begin{pmatrix} 9 - 5 \\ 6 - 3 \\ -2 - 1 \end{pmatrix} = \begin{pmatrix} 4 \\ 3 \\ -3 \end{pmatrix}$

Equation of line through $A$ and $B$:

$\mathbf{r} = \begin{pmatrix} 5 \\ 3 \\ 1 \end{pmatrix} + \alpha \begin{pmatrix} 4 \\ 3 \\ -3 \end{pmatrix}$

$\begin{pmatrix} 5 \\ 3 \\ 1 \end{pmatrix} + \alpha \begin{pmatrix} 4 \\ 3 \\ -3 \end{pmatrix} = \begin{pmatrix} -7 \\ -6 \\ 10 \end{pmatrix}$ is satisfied when $\alpha = -3$ so $\begin{pmatrix} -7 \\ -6 \\ 10 \end{pmatrix}$ lies
on the line.
Hence $A$, $B$ and $C$ are collinear.

## 50 Scalar product

$\overrightarrow{AB} = \begin{pmatrix} -1 - 5 \\ 0 - -2 \\ 6 - 10 \end{pmatrix} = \begin{pmatrix} -6 \\ 2 \\ -4 \end{pmatrix}$

$\overrightarrow{AC} = \begin{pmatrix} 8 - 5 \\ 2 - -2 \\ 1 - 10 \end{pmatrix} = \begin{pmatrix} 3 \\ 4 \\ -9 \end{pmatrix}$

$\cos\theta = \dfrac{\overrightarrow{AB} \cdot \overrightarrow{AC}}{|\overrightarrow{AB}||\overrightarrow{AC}|} = \dfrac{(-6)(3) + (2)(4) + (-4)(-9)}{\sqrt{6^2 + 2^2 + 4^2}\sqrt{3^2 + 4^2 + 9^2}}$
$= 0.3374...$
$\theta = 70.3°$ (1 d.p.)

## 51 Perpendicular vectors

(a) $\overrightarrow{AB} = \begin{pmatrix} 6 - 2 \\ 0 - -1 \\ 3 - 4 \end{pmatrix} = \begin{pmatrix} 4 \\ 1 \\ -1 \end{pmatrix}$

$\mathbf{r} = \begin{pmatrix} 2 \\ -1 \\ 4 \end{pmatrix} + \lambda \begin{pmatrix} 4 \\ 1 \\ -1 \end{pmatrix}$

(b) $\overrightarrow{AC} = \begin{pmatrix} p - 2 \\ -3 - -1 \\ 2p - 4 \end{pmatrix} = \begin{pmatrix} p - 2 \\ -2 \\ 2p - 4 \end{pmatrix}$

$\overrightarrow{AC}$ is perpendicular to $L$ so

$\begin{pmatrix} 4 \\ 1 \\ -1 \end{pmatrix} \cdot \begin{pmatrix} p - 2 \\ -2 \\ 2p - 4 \end{pmatrix} = 0$

$4(p - 2) - 2 - (2p - 4) = 0$
$\qquad\qquad\qquad 2p - 6 = 0$
$\qquad\qquad\qquad\qquad p = 3$

(c) $\overrightarrow{AC}$ is perpendicular to $L$ and $A$ lies on $L$ so perpendicular
distance is:

$|\overrightarrow{AC}| = \left|\begin{pmatrix} 1 \\ -2 \\ 2 \end{pmatrix}\right| = \sqrt{1^2 + 2^2 + 2^2} = \sqrt{9} = 3$

## 52 Solving area problems

1. (a) $\overrightarrow{CA} = \begin{pmatrix} 5 - 6 \\ -1 - -1 \\ 0 - 4 \end{pmatrix} = \begin{pmatrix} -1 \\ 0 \\ -4 \end{pmatrix}$

$\overrightarrow{CB} = \begin{pmatrix} 2 - 6 \\ 4 - -1 \\ 10 - 4 \end{pmatrix} = \begin{pmatrix} -4 \\ 5 \\ 6 \end{pmatrix}$

(b) $|\overrightarrow{CA}| = \sqrt{1^2 + 0^2 + 4^2} = \sqrt{17}$
$|\overrightarrow{CB}| = \sqrt{4^2 + 5^2 + 6^2} = \sqrt{77}$

$\cos(\angle ACB) = \dfrac{\overrightarrow{CA} \cdot \overrightarrow{CB}}{|\overrightarrow{CA}||\overrightarrow{CB}|} = \dfrac{(-1)(-4) + (0)(5) + (-4)(6)}{\sqrt{17}\sqrt{77}}$
$\qquad\qquad\qquad = -0.55278...$
$\angle ACB = 123.5586...°$
Area $= \frac{1}{2}|\overrightarrow{CA}||\overrightarrow{CB}|\sin(\angle ACB)$
$= \frac{1}{2} \times \sqrt{17} \times \sqrt{77}\sin(123.586...) = 15.0748...$
$= 15.07$ (2 d.p.)

(c) $\overrightarrow{OA} - \overrightarrow{CB} = \begin{pmatrix} 5 - -4 \\ -1 - 5 \\ 0 - 6 \end{pmatrix} = \begin{pmatrix} 9 \\ -6 \\ -6 \end{pmatrix}$

$\overrightarrow{OA} + \overrightarrow{CB} = \begin{pmatrix} 5 + -4 \\ -1 + 5 \\ 0 + 6 \end{pmatrix} = \begin{pmatrix} 1 \\ 4 \\ 6 \end{pmatrix}$

$\overrightarrow{OB} - \overrightarrow{CA} = \begin{pmatrix} 2 - -1 \\ 4 - 0 \\ 10 - -4 \end{pmatrix} = \begin{pmatrix} 3 \\ 4 \\ 14 \end{pmatrix}$

Three possible points are $(9, -6, -6)$, $(1, 4, 6)$, $(3, 4, 14)$

(d) Area $= 2 \times 15.0748... = 30.15$ (2 d.p.)

2. (a) $1 + 6\lambda = 5 + 2\mu \qquad ①$
$\qquad\quad -2 = 12 - 2\mu \qquad ②$
From ②: $\mu = 7$
From ①: $1 + 6\lambda = 5 + 2(7)$
$\qquad\qquad\quad 6\lambda = 18$
$\qquad\qquad\qquad \lambda = 3$

$\begin{pmatrix} 1 \\ -2 \\ 4 \end{pmatrix} + 3\begin{pmatrix} 6 \\ 0 \\ 1 \end{pmatrix} = \begin{pmatrix} 19 \\ -2 \\ 7 \end{pmatrix}$

$P$ is $(19, -2, 7)$

(b) $Q$ is $(1, -2, 4)$
$R$ is $(15, 2, 5)$

$\overrightarrow{PQ} = \begin{pmatrix} 1 - 19 \\ -2 - -2 \\ 4 - 7 \end{pmatrix} = \begin{pmatrix} -18 \\ 0 \\ -3 \end{pmatrix}$

$\overrightarrow{PR} = \begin{pmatrix} 15 - 19 \\ 2 - -2 \\ 5 - 7 \end{pmatrix} = \begin{pmatrix} -4 \\ 4 \\ -2 \end{pmatrix}$

$|\overrightarrow{PQ}| = \sqrt{18^2 + 0^2 + 3^2} = \sqrt{333}$
$|\overrightarrow{PR}| = \sqrt{4^2 + 4^2 + 2^2} = \sqrt{36}$

$\cos(\angle QPR) = \dfrac{\overrightarrow{PQ} \cdot \overrightarrow{PR}}{|\overrightarrow{PQ}||\overrightarrow{PR}|} = \dfrac{(-18)(-4) + (0)(4) + (-3)(-2)}{\sqrt{333}\sqrt{36}}$
$\qquad\qquad\qquad = 0.7123...$
$\angle QPR = 44.5698...° = 44.57°$ (2 d.p.)

(c) Area $= \frac{1}{2}|\overrightarrow{PQ}||\overrightarrow{PR}|\sin(\angle QPR)$
$= \frac{1}{2} \times \sqrt{333} \times \sqrt{36}\sin(44.5698...) = 38.42$ (2 d.p.)